Tennis Science
for
Tennis Players

Tennis Science
for
Tennis Players

Howard Brody

u//

University of Pennsylvania Press
Philadelphia

To all my girls

Copyright © 1987 by the
University of Pennsylvania Press
All rights reserved

Library of Congress Cataloging-in-Publication Data

Brody, Howard, 1932–
    Tennis science for tennis players.

    Includes index.
    1. Tennis.   2. Tennis—Equipment and supplies.
I. Title.
GV995.B6923   1987   688.7'6342   86-30735
ISBN 0-8122-1238-X (pbk.)

Printed in the United States of America

Line figures drawn by Joseph Guerrero, Jr.

Fifth paperback printing 1992

Contents

Tennis Science
for
Tennis Players

Introduction
Why Another Book on Tennis?

This book will not make you an instant champion; it will help you to play as well as your physical endowment allows. There is no substitute for athletic ability and practice, but there are many things that will help you to win more points—without the drudgery of long hours of lessons, practice, and hard work. In addition, taking an understanding of all the laws of nature onto the tennis court with you will add to your enjoyment of the game.

You do not need knowledge of science or engineering to use the material presented here. In order to profit from this book, it is not necessary that you understand, or even read, the explanations of the physics. You are free simply to accept what is recommended here, just as you do when you learn tennis or take lessons. You never ask the pro why you should keep a firm wrist or why you must follow through with every stroke.

Written for tennis players, this book is based on work that has been done in the laboratory and on a computer at the University of Pennsylvania over a number of years. Some of the information has been published in four rather technical articles for physicists in the *American Journal of Physics* (June 1979; September 1981) and *The Physics Teacher* (November 1984; April 1985).

This book gives advice on three aspects of the game that determine and shape one another: equipment, strokes, and strategy. Most important is the matching of these three things to your own ability. Because Bjorn Borg strings his racket at 80 pounds, you should not necessarily do likewise. Because Jimmy Connors throws his whole body into every shot, you are not required to do so also. And because some top pro blasts his first serve and eases up on his second, you need not feel compelled to do the same. All the best professional players have tuned their equipment, developed their strokes, and calculated their strategy to fit their own physical abilities and temperaments—not yours—so you should not blindly copy them.

This book lays no claim to being The Compleat Handbook of Tennis. It covers only a limited number of topics, but it does so from a different point of view—that of a scientist trying to tell you how to take advantage of the laws of nature to win more points and, as player or spectator, to enjoy the game more. Tennis has developed through years of trial and error. No group of scientists, engineers, and players has ever attempted to determine the underlying scientific principles of the game and then to construct strokes and strategies based on their analyses. Professional teachers and books on tennis answer all the "how to" questions, but they do not try to answer the "why" questions (and they certainly do not bring the "how to" and "why" together, as here). The pros have learned tennis by taking lessons from other pros or coaches, and a set method of teaching tennis has been developed. It is basically an arbitrary series of rules and drills, and it works. Up to now, very few people have stopped to think about the basic science that governs the game, and no one has yet published a book of this type.

Strokes are not emphasized in this book. Almost every book on tennis will tell you more than you probably want to know about how to hit a backhand, how to manage the mechanics of the overhead. Instead, this book stresses the proper choice of equipment, strategy, and ball trajectory because scientific analysis of these subjects gives meaningful results that can be translated into specific actions that you, the tennis player, can take advantage of and appreciate. Several additional topics are included, even though they lead to no specific recommendations, because they are appropriate and interesting, and because the information cannot be found anywhere in the tennis literature.

A few of the concepts advocated in this book are contrary to accepted tennis lore. These new ideas are not arbitrarily fashioned; they are derived from the application of the basic laws of physics to the game of tennis. In addition, many of them have been tested in the laboratory, with computer models, and, of course, wherever possible, on the court. Ten years ago, *many* of the ideas in this book would have been dismissed by the tennis establishment as wrong. No one had done controlled experiments, and the accepted lore was based on anecdotal information. There was no doubt in most players' minds that tighter strings and a flexible racket gave more power. Today we know better. With the revolution in tennis racket technology and design that began ten years ago, manufacturers and designers of rackets developed a new understanding of tennis. Now it is necessary that players and tennis teachers also come to a new understanding of the game. That is what this book is all about.

Chapter 1
The Importance of the Strings

There is a famous advertisement in many tennis magazines that points out that the ball never touches that very expensive, high-tech tennis racket. The ball only touches the strings. That is why this book will start by discussing the strings.

1.1 Strings and Stringing

What tension should you string your racket at? Should it be 50 pounds? Or 55? Or 62? Or 73? How can you determine what tension is optimum for a racket, a style of play, and the strings that are in use? When players decide to have an old racket restrung—or even to get a new one—they usually ask for the same tension that they have been using. If they are at all unhappy about their present game, they may think about trying a different tension. Unfortunately, if they ask someone in a sports store, a teaching professional, a friend, or a good player what tension they should try, they will usually not get good advice because most people do not understand the physics of the strings.

Some of the Problems

There is more to the strings and stringing than just the tension. Not long ago, when a racket was strung, the head sizes were all the same, the string spacing was uniform, and specifying the tension was about all one had to do. Now there are a variety of head sizes and stringing patterns. A tension of 65 pounds in a standard size racket plays very tightly, while 65 pounds in an oversize frame may play loosely. How do you compare rackets with different size heads? How do you compare rackets where the space between the strings (string density) is different or nonuniform? You can buy rackets where the strings are twice as far apart as in the standard racket, and there are rackets where the spacing is half that of the standard. In addition there are rackets with three sets of strings instead of the familiar two. When you use a different gauge of string, how does that change the way the racket plays? What is the difference between gut and synthetic strings?

Some Answers

The way the racket plays, with respect to the strings, can be determined by examining how much the string plane deforms when a force is applied to the racket face. If you push on the strings with a known force and measure the deformation of the string plane (how much it moves perpendicular to the plane of the strings), you will know approximately how the racket will play. This measurement automatically takes into account the size of the racket head, the density of the stringing pattern, the tension, and most of the other variables that can be involved. You can either take a single measurement with a single force pushing on the strings, or you can measure the string deformation for many values of the force, as is shown in Figure 1.1.

GENERAL RULE I:
▶ Rackets will play in a similar manner if they are strung so that their curves of string plane deformation versus force are similar.

By measuring the string plane deformation or deflection, therefore, you can compare a Prince (oversize racket) strung with 15-gauge nylon with a Wilson Kramer strung with 16-gauge gut and know how the strings in one will play relative to the other. Some tennis shops have a device to do this (for example, the Sports Pal "Flex II" Tension Tester), or you can do it yourself (as was done to obtain the data used in this book) with a set of weights and an accurate ruler. Figure 1.1 shows several different rackets tested this way. It is obvious that the strings in the old Spalding Smasher strung at 20 pounds of tension deflect a great deal more (for the same applied force) than, for example, those of the Prince, which was strung at 76 pounds of tension.

GENERAL RULE II:
▶ If you increase (or decrease) the tension of the strings in proportion to changes in the length of the strings in the head, the string plane deformation will be similar to first order.

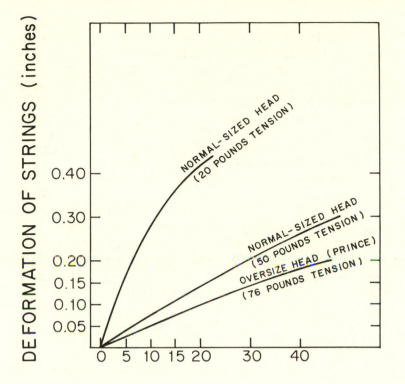

FORCE AT CENTER OF HEAD (pounds)

Figure 1.1. The String Plane Deformation versus Applied Force for Several Rackets. As the force being applied perpendicular to the face of the strings increases, the deformation of the strings increases. The greater the tension (divided by string length), the less the strings will deform for a given applied force. This figure shows rackets with string tensions ranging from 20 pounds to 76 pounds.

As an example, an 8-inch-wide head (which is the old standard size racket), strung at a tension of 55 pounds, will have a string plane deformation similar to that of an oversize 10-inch-wide frame (which is 25 percent larger), strung at 69 pounds (25 percent greater), if all other things (such as string density, gauge, and so on) are the same. This means that in order to change from one frame size to another while retaining similar playing characteristics from the strings, the tension divided by string length must be kept the same. This is why the oversize racket is strung at higher tensions.

For small string plane deformations (hitting the ball softly), the principal variables that determine the string response are the tension divided by string length and the density of the stringing pattern (number of strings per inch in the weave). For moderate hits (larger string plane deformations), the elasticity of the strings begins to affect the way the racket plays. The harder the ball is hit, the more important the elasticity of the strings becomes. The more elastic the strings are, the larger the string plane will deform for a given applied force, and the straighter the curve of string plane deformation versus applied force will be (the less the curve will turn over or level off at large forces).

1.2 Power from the Strings

The reason you go to the trouble of using a strung racket instead of simply a wooden paddle is so that you can get power. You want the ball to leave the strings with a high velocity without your having to swing the racket at a very high speed. Strings allow you to do this. The tighter you string your racket, the more it feels like a wooden board, and the less power you will get. That last statement bears repeating:

▶ Tighter strings mean less power; looser strings mean more power.

The physics of this is clear and easy to understand. Then why do so many people think just the opposite? They see many of the top players get tremendous power from their rackets, which are strung tightly. Borg strings his racket at 78 pounds, so they conclude this is why his balls have great pace on them. Nothing could be further from the truth. It is because Borg and the other top players can hit the ball so hard that they can afford to string their rackets tightly to gain other advantages; they sacrifice some of their power in doing so.

Why Loose Strings Give More Power

By design, a tennis ball does not store and return energy efficiently. A ball dropped from a height of 100 inches onto a hard surface rebounds only to a height of about 55 inches. Indeed, that is the official specification for the manufacture of a tennis ball. This means that about 45 percent of the energy that the ball had has been lost ($100 - 55 = 45$).

Strings, on the other hand, are designed to return between 90 and 95 percent of the energy that is fed into them. To give the ball the maximum energy (that is, the highest speed) when it is hit, the strings, not the ball, must store the energy by deflecting. The strings return almost all of the energy that they store when they deform, while the ball only returns about half the energy when it is hit and deforms. If the strings have a lower tension, they will deflect more (that is, store more energy), and the ball will deform less (that is, dissipate less energy). The larger the string plane deformation, the less the ball will deform. Try dropping the ball on the ground (equivalent to infinitely tight strings) and comparing the rebound height to that when the ball is dropped from the same height onto a racket lying on the ground (supported around the head, not by the handle alone). The ball will bounce back considerably higher in this second case—to 80 percent of its original height, compared to 55 percent when it hits a hard surface. This 25 percent additional bounce shows that less energy is dissipated when the ball hits the flexible strings than when it hits the inflexible floor (infinitely tight strings). In addition, laboratory tests with high-speed equipment show that with looser strings the ball's rebound velocity is greater.

Of course, there is a limit to getting more power by loosening the strings—you cannot play tennis with a butterfly net. Once the string tension is reduced so much that the strings begin to shift their position in the plane of the racket head and rub against each other, energy is lost. If you reduce the tension too much, you will not get the ball speed that you want, and the strings will wear out too fast from the excessive rubbing.

Why Not String All Rackets Loosely?

If all you wanted from a racket was power, then the conclusion is clear: *String loosely*. But there is something else that you want from a racket, and that is control. You want to be able to put the ball in the court, and often you want to be able to place the ball down the line or at a sharp angle. You want to be able to control the pace of the ball and where it goes, even if you do not hit it perfectly on every shot. There seems to be a consensus that reducing the tension in the strings of a racket means sacrificing some control, although this has not been proven. It is very difficult to do tests in the laboratory or on the court to measure control because the results are not reproducible. (It is much easier to study ball speed, hence power, because measurements in the laboratory are reproducible.) A number of reasons have been put forward to explain why looser strings tend to harm your control; they are listed here but not commented on.

1. Stringing the racket at low tension leads to what is called the "trampoline" or "slingshot" effect. The ball rockets off of the racket and therefore is harder to control.

2. Stringing at lower tension tends to make the speed of the ball as it leaves the racket more dependent upon the pace of the opponent's shot; hence control is somewhat reduced.

3. If the ball hits the racket face off-center, looser strings will tend to change the angle at which the ball leaves the racket more than tighter strings will.

4. Looser stringing increases the dwell time of the ball on the strings (how long the ball remains in contact with the strings). This allows the racket to twist or turn more while the ball is still in contact; hence the ball leaves the racket at a greater angle.

5. Tighter strings bite into the ball more and allow greater control.

6. The ball flattens out more on tighter strings; because there is more contact area, there is more control.

7. Tighter strings have a more linear response when the ball is hit harder.

There is, however, an argument that control increases with looser strings if the swing is modified to take advantage of the extra power the loose strings give. If the ball will leave the racket with a higher speed, then it is not necessary to swing the racket so hard and take so full a swing. This reduction in the needed racket speed can lead to greater control over the racket position, therefore giving better control of where the ball goes.

If you do not swing your racket hard, you will often benefit from looser strings. They will tend to convert your opponent's power to your power with high efficiency. If you really want to swing out at the ball, then use a tightly strung racket. You will get that extra bit of control that hard hitters always seem to need, and the ball will still have a reasonable pace. And if you want to hit shots with a great deal of spin on them, then tight strings are an advantage. But remember, if you love to hit the ball really hard and you have a racket that is strung loosely, then you will probably win awards for the longest drive off the tee, not for tennis matches. If, on the other hand, you have a very compact swing, or you tend to punch out at the ball rather than swing, then loose stringing may help your game as it does for John McEnroe.

1.3 Loose versus Tight Strings and Dwell Time

The looser the strings (actually the greater the string plane deformation for a given force), the longer the ball will reside on the strings (the greater dwell time). The dwell time of the ball on the strings should increase as the inverse of the square root of the tension; measurements made in the lab bear this out. In addition, the dwell time of the ball on the strings decreases the harder the ball is hit, because the strings become effectively stiffer the more they are forced to deform. This last statement may seem contrary to common sense, if you picture a high-speed tennis ball sinking into the strings more than a gently hit ball and therefore conclude that the dwell time should be greater for the ball that is hit hard. It is nevertheless correct, since the strings get stiffer nonlinearly the more they deform; this leads to shorter dwell times for harder hits.

Measurements of these effects, taken with a laser and high-speed electronic equipment, are summarized in Figure 1.2. This figure shows, for many different ball speeds, the dwell time of tennis balls on rackets with different string tensions. From these data it should be clear that if someone hits a ball hard to you and you try to return it hard, the dwell time of the ball on the strings will be less than if you play a soft game. If you decrease the tension of the strings, the ball will spend more time in contact with them; as you increase the tension, the dwell times will decrease.

How does the dwell time affect your tennis? The actual time of contact for a normal shot with normal string tension is about 4 or 5 thousandths of a second. By reducing the tension and not hitting hard you might be able to increase the dwell time to about 6 or 7 thousandths of a second, but you have not accomplished

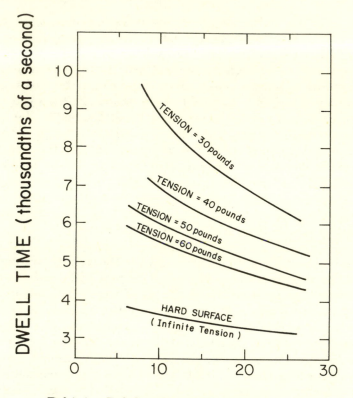

Figure 1.2. Dwell Time of the Ball on the Strings. The dwell time of balls hitting rackets with different tensions is shown here for a range of ball speeds. The looser the strings, the longer the dwell time. The tighter the strings, the shorter the dwell time. This figure shows that the higher the ball speed, the shorter the dwell time.

much by gaining that extra 2 or 3 thousandths of a second. These times are so much shorter than your reaction or reflex time that you cannot possibly do anything to change the shot while the ball is on the strings. In short, although a shot may not feel right, by the time that bit of information is registered in your brain, the ball is well on its way back toward your opponent.

A longer dwell time corresponds to a longer "carry distance" (that is, the distance the ball and the racket move together while the two are in contact). An excessively long dwell time can lead to a loss of control if the racket head changes its angle relative to the court while the ball is being carried. To some degree this happens on all shots, but if the change in angle is appreciable owing to an off-center hit (the racket twists), then a loss of control could result.

A longer dwell time also means that the shock of the ball being hit is spread over a longer time; the magnitude of the force at any given time is therefore reduced. In other words, when the strings are softer (at lower string tension), the arm does not feel the shock of hitting the ball as much as when the strings are harder (at higher string tension). If you wish to alleviate arm troubles, such as tennis elbow, reducing the tension of your racket strings will not only lessen the initial shock transmitted to your arm but will give you more power from the racket. You will not have to swing so hard, which is kinder to your arm. This may not cure your tennis elbow, but it clearly is a step in the right direction.

A lot of confusion still prevails on the subject of power versus string tension. If the ball spends less time on tighter strings, doesn't that mean it leaves the racket quicker, hence with higher speed? This seems to contradict the above statements that lower string tension gives more power to the ball. True, the ball does leave the tighter strings "faster," but this merely means that the ball spends less time in contact with the strings. In other words, although the ball leaves the strings sooner when the strings are tight than when they are loose, that does not mean the ball is moving faster when it leaves the tighter strings. The exact opposite is true.

One of the stock phrases many teaching professionals and tennis books use is, "for better control, keep the ball on the racket as long as possible." Statements like this appear also in advertisements for tennis balls and in the patent literature. It is clear that all one need do is reduce the tension in the strings to ensure that the ball will spend more time on the racket. Yet, as has been discussed, loose strings decrease rather than increase a player's ability to control the ball. Clearly, something is wrong with the advice being given. What tennis teachers should say is, "Swing your racket so that it appears you are maximizing the time that the ball spends on the racket." This makes sense, does not contradict what we know from the physics of strings, and gives the student good strokes.

1.4 Stringing Material

What is the best kind of string to have in your racket, gut or some kind of synthetic string? It is generally accepted by experienced tennis players that gut is superior to nylon in its "playing characteristics," but it is difficult to find specific reasons for this superiority. Technical data from the manufacturers of strings contain the same words and phrases that are used in their advertisements. Everyone agrees that gut is more expensive, more fragile, less durable, and less resistant to damage by moisture and humidity than synthetics. What wonderful characteristics could gut have in play that would offset all these reasons against using it?

The elongation of a piece of gut string and a piece of nylon string has been measured as force was applied to each. The data are plotted in Figure 1.3. The important parts of the curves are those between 50 and 75 pounds of tension—the range over which rackets are normally strung. In this range of tensions, the nylon string requires 12 pounds of additional tension to elongate it an extra 1 percent while the gut string needs only 6 pounds of additional tension to stretch it an extra 1 percent. What is important is the *additional* tension needed to stretch the string an *extra* one percent, and not how much the string stretched to get to the 60 or 65 pounds of tension used during play. This means that when a given force (for example, a ball hitting the strings) is applied, the gut strings will stretch more than the nylon strings of the same diameter (gauge), provided each set of strings is strung to the same tension. Because of the greater elasticity of gut, it will deform more and tend to cup the ball.

This extra power that gut gives you is a function of how you hit the ball. If you play a soft game, the advantage of gut is reduced. If, on the other hand, you really smash the ball, then you will appreciate the extra power that gut provides. In addition, the greater flexibility of gut may give you more control over the ball, because you will get a more linear response from the strings as you vary your swing and the ball speed. Gut will also be somewhat kinder to your arm than nylon because of its increased elasticity.

These beneficial effects can be enhanced by using a thinner gut such as 16L gauge (between 16 and 17 gauge), or even a very thin string such as 17 gauge. Clearly, thinner strings are more fragile, but the best players find that this disadvantage is outweighed by the benefits they provide. For the ordinary player who does not hit the ball hard, however, the extra money spent in buying gut or using a higher gauge string is probably wasted.

Gut is somewhat rougher in texture than most synthetic fibers, so in addition to its cupping action, it may tend to grip or bite into the ball a little better than nylon, leading to a better

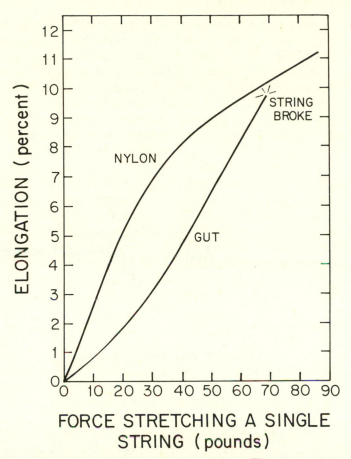

Figure 1.3. Gut versus Nylon under Stretch Test. This figure shows the elongation of a single string of gut and a single string of nylon when subjected to a stretching force. At 60 to 70 pounds tension the gut line is curving upward and the nylon line is becoming more level. This means that gut will be more elastic at these tensions. (This data was taken using 15L-gauge strings.)

touch or feel on shots. There are some rough-textured synthetic strings on the market that attempt to duplicate this feature of gut, but it is not clear how well they work. The weaving of the strings over and under each other provides a roughness that is much greater than that inherent in the strings themselves.

When the large-headed rackets first became available, most of the gut on the market would break at the higher tensions needed to make these rackets play well. Therefore, they were initially all strung with synthetics. When put into the oversize head, the synthetic strings did not have some of the harsher properties that some players find in conventional-sized rackets strung with nylon. Computer simulations of the strings in normal and larger racket frames show that when the oversize racket is strung with a typical synthetic, it has many of the properties of a conventional racket strung with gut. Recently gut has been produced that will withstand the tensions necessary for the oversize rackets.

There are on the market all types of strings called "synthetic gut" that claim to have the playing characteristics associated with gut. They may have, but playing characteristics are determined by each individual player. Polyester tests very favorably in the laboratory, but players who have used it on the court do not seem to be impressed. In one test that compared rackets strung with gut and rackets strung with synthetic strings, a surprisingly large number of players could not distinguish the two. It is even more difficult to tell them apart in the large-headed rackets.

1.5 String Thickness

The gauge of a tennis string corresponds to the standard U.S. wire gauge thickness (Table 1.1). This means that 15 gauge is 12 percent thicker than 16 gauge, and 17 gauge is 11 percent thinner than 16 gauge string; 15L and 16L are between 15 and 16, and 16 and 17 respectively.

If you want to string or restring a racket yourself, you should have a device for measuring the thickness of the new strings and those that are already in the racket. You can use a fixed template with a series of notches in it, or you can use a micrometer, which is a device that can measure the diameter of a string to a ten thousandth of an inch. They are available calibrated in millimeters or in mils (thousandths of an inch); there are 25.4 millimeters in an inch and about 39.37 mils in a millimeter, if you want to convert from one unit to the other.

Table 1.1. Wire Gauge Thickness

Gauge number	Thickness		
	mm	inch	mil
15	1.45	.0571	57.1
16	1.29	.0508	50.8
17	1.15	.0453	45.3

When you measure a string with a micrometer, you probably will find the measurement does not correspond exactly with one of the values listed in Table 1.1. You should expect some variation from string to string, and between different materials and different brands of string. If you measure a string under tension (in the racket for example), you will find that it is thinner than it was before it was put into the racket. That is because the string has stretched, and in order to do this it must get thinner (a conservation law of physics). If the string has stretched 10 percent, then it should be about 5 percent thinner. (A 50-mil, 16-gauge string that is stretched so that it is 10 percent longer will become only 47.5 mils thick, or the diameter of a 16L string.) There is a general rule: If a string stretches a certain percentage, its thickness will decrease by about half of that percentage.

But how does the thickness of the string affect the way it plays in a racket? It turns out that it is not the string's thickness but its cross-sectional area that is of direct interest to a tennis player. The area of the cross section determines such properties of the string as its strength, elasticity, and stretch.

The area is proportional to the square of the thickness (it is actually equal to π over 4 times the thickness squared), so changing from 16- to 15-gauge string (an increase in thickness of 11 percent) will increase the cross section by the square of 11 percent, which is about 23 percent. As a general rule, if the string thickness is increased by a certain percentage, the cross-sectional area will be increased by approximately twice that percentage. Because of this increase of 23 percent in the cross-sectional area, for the same type of string material, 15-gauge string will be 23 percent stronger, will stretch 23 percent less (under the same tension), and will be 23 percent less elastic than 16-gauge string.

This last effect, the change in the elasticity, is of great importance. It is the elasticity of the strings that makes your racket play well. Gut is more elastic (for the same diameter or gauge of string) than synthetics, which is why many people prefer gut. When you use thinner strings, you also gain elasticity. Since a 17-gauge string is 20 percent thinner than a 15-gauge string, the cross section will be about 40 percent smaller. Therefore, 17-gauge strings will be almost twice as elastic as 15-gauge strings. The 17-gauge strings will not stand the same tension, will not wear as well, and will pop more often, however. It is a case of durability versus playability. If money is no object, and you are willing to bring a large number of rackets whenever you play, then 17-gauge strings could be for you.

1.6 Frequency or "Ping" of the Strings

Many stringers and players check the tension in a racket by the sound of the strings when they are plucked. This is a fairly accurate test only as long as the string gauge, the weight of the material in the string, and the size of the racket head are always the same.

If you calculate the physics of the string vibration, you will find that the frequency of the oscillation (the pitch of the sound) is proportional to the square root of the tension, proportional to one over the string length, and proportional to one over the string thickness. For example:

> If the head size and string gauge are fixed and you raise the tension, the frequency (pitch) will go up.
> If the string tension and gauge are kept constant, a larger head (hence longer strings) will give lower pitches.
> If string tension and head size remain the same, thicker strings (lower gauge) will produce lower pitches.

This means that you can be fooled by the pitch of a string. If you have two rackets, identical except for different gauge strings, and you adjust the tension to get the same sound in both, the thinner string will have a lower tension. If you string an oversize racket to sound the same as a standard racket, the tension will be much higher than you want. Since General Rule II said to increase the tension in proportion to the increase in length of the string in order to achieve comparable playability, when using a bigger racket you would get a lower frequency sound, even though the tension is higher.

1.7 Tension versus Time

If you had your racket strung at a tight 60 pounds several years ago, you would probably guess that the tension had relaxed—maybe down to 55 pounds or so. Actually, the tension might be considerably lower, depending upon the type of string material, the gauge, and the frequency of use. If it were now 50 pounds, you might not have the racket restrung, having just read that lower tension leads to more power. This line of reasoning may be flawed, however, in that it ignores another significant fact.

The loss of tension over time results from a stretching of the strings; that is, there is a gradual breaking or sliding of the bonds in the molecules, especially if the stringing material is a synthetic like nylon. There may be a slight loss of the elasticity of the material as well. If the material becomes less elastic it will not give back to the ball exactly as much energy as it did when it was new. Therefore, strings that stretch and become loose may not be as resilient as strings that are strung at lower tension and are fairly new. It is possible you will have less power with the old strings than with new strings at a low tension. In addition, the old looser strings will contribute more to a loss of control over the ball compared to new strings at lower tension because of the loss of elasticity.

If you were to hang a 60-pound weight on a single string and measure the length of the string, you would discover that the string's length would increase quite a bit (beyond the initial stretch) during the first hour, not quite so much during the second hour, and so on until after hundreds of hours the stretching would seem almost to stop. (The string would actually continue to increase in length almost logarithmically in time, when kept at constant tension.) Measurements of the stretch of several types of strings are shown in Figure 1.4. When strung in a racket, where it is kept at constant length but not constant tension, a piece of string will approach an equilibrium tension in less time than it would in the test described above. This is because the tension is not constant but rather decreases as the string stretches. As a result, most of the relaxation due to stretching probably takes place before the racket is used or within the first few days. An additional loss of tension will take place during the first few hours of play with the racket. The additional stretching in the racket over subsequent months will be much less than the elongation of the string would be under conditions of constant tension. If you measure the tension in your racket every month or two, however, you will probably notice some deterioration. Measurements of the string plane deformation of a racket were performed several times over the course of a few weeks; the results are shown in Figure 1.5.

1.8 The Density of the Stringing Pattern

If the strings in a racket are spaced a half inch apart, the racket will play differently than a racket with ¾-inch spacing, even if the tension, racket size, gauge, and so on are the same. A denser pattern (more strings to the inch) will play stiffer, or as if the tension were higher. If a racket has twice as many strings to the inch,

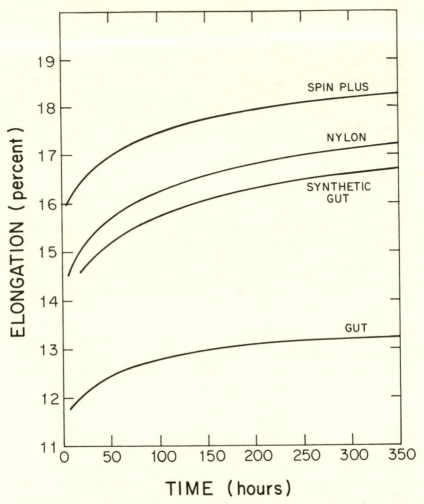

Figure 1.4. String Elongation versus Time. Lead weights were hung on single lengths of several types of string for a period of many hours and the percentage of the elongation (stretch) was recorded.

then it should be strung at half the tension to give it the same string plane deformation and consequently the same feel. It will then be possible to use a thinner string, which is more elastic and therefore perhaps more desirable. If the strings are farther apart, the string plane deformation will be larger and the racket will play as if the tension were lower. A racket with half as many strings to the inch should be strung at twice the tension, but to handle this higher tension, a thicker, and thus less elastic, string

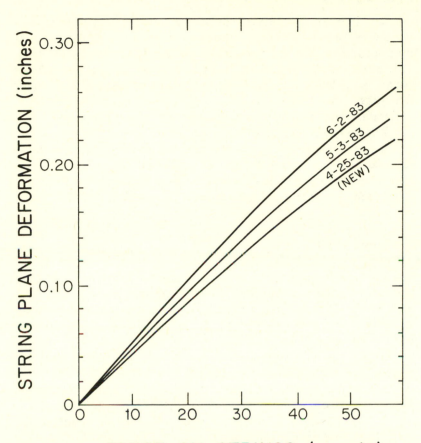

Figure 1.5. String Plane Deformation versus Time. The string plane deformation of a single newly strung racket was measured at three different times to show how the relaxation (or creep or elongation) of a string shown in Figure 1.4 affects the tension when that string is used in a racket. The increase in deformation (for the same force) means that the strings are at a lower tension.

must be used. Hitting a ball under these conditions may give the player a harsher, stiffer feeling from the strings.

Racket manufacturers make use of this effect of string density when the string pattern of a racket is designed. In a uniformly strung racket head, the strings are normally "softest" in the center; they get effectively stiffer the closer they are to the frame. By increasing the string spacing close to the frame, racket designers try to soften the strings away from the center. Consequently, many rackets have a dense stringing pattern in the center and a somewhat sparser pattern near the periphery in an attempt to increase the power from the strings in the latter area. The same result can be accomplished by stringing the shorter strings at a lower tension than the longer strings. This is more difficult for the stringer, requires a string lock in each frame hole, and means that two strings cannot use the same hole.

The Bergelin Longstring racket, in an attempt to spread the power over its entire face, has strings at 45 degrees; the spacing is close together near the throat and increases with distance from that area. A device inside the handle allows the player to adjust the tension in all the strings, right on the court.

1.9 Summary

An increase in string tension will decrease the power, increase the control, decrease the dwell time, and increase the shock to your hand. Conversely, a decrease in string tension will increase the power, decrease the control, increase the dwell time, and decrease the shock to your hand.

As you change racket size, you should increase or decrease tension in proportion to string length.

More elastic strings play better. Gut is more elastic than synthetics, and thinner strings are more elastic than thick strings.

A larger spacing between strings will make the strings play as if they were at a lower tension. Conversely, a smaller spacing between strings will make strings play as if they were at a higher tension.

Most strings lose tension with time, and this decrease in tension can be accompanied by a decrease in elasticity.

Chapter 2
The Sweet Spots of a Tennis Racket

When you hit a shot and it really feels good, you claim that you have hit the sweet spot. But can this feeling be quantified? What is the real definition of sweet spot? When you hit a ball in the wrong place on the racket, you experience any one of several responses. A shock may be transmitted to your arm, the whole racket may seem to shudder and vibrate, or the ball may not leave the strings with the speed or power that you expect. What causes these shock tremors, vibrations, and losses of power? Do some rackets have a sweet spot that is sweeter than that of other rackets, or is the size of the sweet spot the only relevant consideration?

Any tennis magazine has advertisements for the various rackets in which each manufacturer claims the biggest or best sweet spot. Even if the manufacturer actually measures the size or quality or location of the sweet spot with electronic equipment, does this have anything to do with how the racket plays or your own experience on court? This chapter will try to answer some of these questions and explain what the manufacturers and their advertising agencies are saying about rackets.

2.1 The Sweet Spot Trio

Even though the term *sweet spot* has existed for a long time, it was not defined in the scientific literature until a technical paper was published in 1981 in the *American Journal of Physics*. This article pointed out that there are actually three sweet spots in a racket, each of which measures a different physical characteristic of the tennis racket. Therefore, your confusion and many of the racket manufacturers' seemingly conflicting claims may all be justified. The three sweet spots are defined as those places where, when you hit the ball, you experience one of three things.

*Sweet Spot 1: The initial shock to your hand is at a
   minimum.*
*Sweet Spot 2: The uncomfortable vibration that your hand
   and arm feel is at a minimum.*
*Sweet Spot 3: The ball rebounds from the strings with
   maximum speed (or power).*

Now, in the language of engineers and scientists, each of these
spots has a special, technical name:

Sweet Spot 1 is the center of percussion (COP).
Sweet Spot 2 is the node of the first harmonic (the node).
*Sweet Spot 3 is the maximum coefficient of restitution
   (COR).*

In general, these three spots are found at separate locations
on the face of the racket. Each spot is a defined region rather than
an actual point, and it is determined completely arbitrarily by the
person doing the testing. For example, the second sweet spot is
the region in which the vibration caused by the ball hitting the
racket is less than some arbitrary value. The third sweet spot is
the region in which the ball rebounds above some arbitrary value.
The actual points at which vibrations are at a *minimum* or at
which the *greatest* ball return speed is obtained are not arbitrary.
They can be determined by experiment. Typical locations for
these three spots are shown in Figure 2.1. The measurable locations of these points are caused by several factors working together: the racket's weight distribution, its flex, its head size and
shape, and so on. Racket designers, of course, know that adding
weight to the tip of the racket or changing the flex of the head will
move the sweet spots. In short, the locations of the sweet spots
can, and do, vary. It is probably true that no racket available today has all three of its sweet spots at the center of the strung area.
An ad that touts the sweet spot rarely states which sweet spot is
being praised; of course it will be the one that makes that racket
look better than its competition.
    Can racket users determine the location of each of these
sweet spots on their own? Yes, and the simple tests described below do not require a well-equipped physics laboratory; they can be
done at home or in an ordinary workshop.

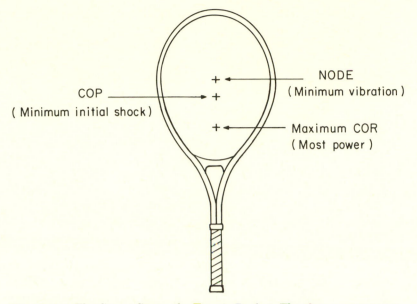

Figure 2.1. The Sweet Spots of a Tennis Racket. The three sweet spots of a tennis racket are shown in their typical positions on the face of a tennis racket.

2.2 Sweet Spot 1: Minimum Initial Shock to Your Hand (The Center of Percussion)

When an unrestrained racket is struck by a ball, the racket re-coils to conserve momentum. If the ball were to hit the racket at its center of mass (CM) or balance point (which is usually in the throat of the racket), the racket recoil would be pure translation and there would be no rotation of the racket. Instead, if the ball were to hit in the center of the strung area, the racket would both translate (to conserve linear momentum) and rotate (to conserve angular momentum), as is shown in Figure 2.2. There is then one point in the handle of the racket where the motion due to the translation (which is in the direction the ball was moving before it hit the racket) and the motion due to the rotation (in the direction of the ball rebound) cancel each other. If you were holding the racket at exactly that point, your hand would feel no shock or impact when the ball hit the racket because the racket would tend to pivot about that point and not move. The location of the point where the ball hit would then be called the center of percussion (COP), but this point can only be determined once the hand loca-

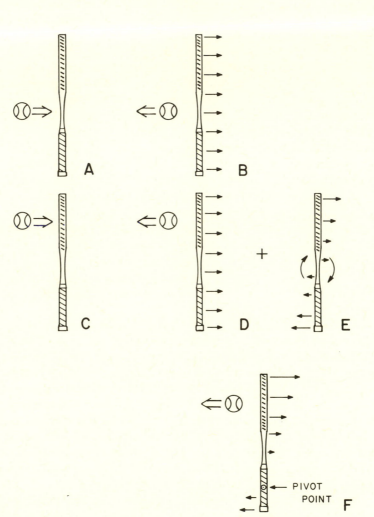

Figure 2.2. Center of Percussion. When a tennis ball hits a racket at the center of gravity (A), the racket will rebound and will translate but not rotate (B). When a tennis ball hits a racket at the center of the head (C), the racket's motion will be both a translation (D) and a rotation (E). Combining the rotation and the translation into a single motion (F), shows that there may be one point in the racket that does not move and the racket rotates about that point. If you were holding the racket at that point, your hand would feel no initial shock or jar when you hit the ball. The location of this pivot point will vary as the ball hits different points on the racket. When the pivot point is the point at which you are holding the racket, the ball has hit the center of percussion of the racket.

tion (or pivot point) is established. If you hold your hand a bit farther up the racket, or down very close to the butt end, the COP will be at a slightly different location on your racket, since it is defined relative to where *you* hold your racket.

When you hit the ball at Sweet Spot 1 (the center of percussion), the initial shock or force on your hand is at a minimum and the shot feels good. The point in the racket handle that you are holding (under the base of your index finger) will not move relative to your hand. If you do not hit the ball at exactly Sweet Spot 1, there will be an initial net force on your hand. For example, if you hit the ball at a point on the head farther away from your hand than the COP sweet spot, the racket will try to pull itself out of your hand (on your forehand shots) by opening your fingers. If you hit the ball closer to your hand than this sweet spot (which is difficult to do with a standard size racket), the initial force will push on the palm of your hand.

Locating the point that minimizes the initial shock to your hand (finding the center of percussion)
Hold the racket between your index finger and your thumb at the place on the handle where you normally put the base of your index finger (about 4 inches from the butt end). Allow the racket to swing as a pendulum with your fingers as the pivot and your hand and arm not moving. Place your fingers so that the racket swings with its face open, as if it were hitting a low ball. Then, measure the time (in seconds) it takes for a complete back-and-forth swing of the racket. Better still, allow the racket to swing back and forth from the starting point five times, record the total time elapsed, and divide this number by five. This will give you a more accurate value of the time of a single swing (t/swg).

The distance (in inches) from where you are holding the racket handle to the minimum initial-shock sweet spot is found by multiplying the time per swing squared (t/swg × t/swg) by 9.77.

$$\text{Location of COP (in inches)} = 9.77 \times (\text{t/swg}) \times (\text{t/swg})$$

For example, if the time of one complete swing is 1.3 seconds, then the distance from your fingers to Sweet Spot 1 is 16.5 inches, which is (9.77 × 1.3 × 1.3). This is almost exactly how the location of the center of percussion is determined in the laboratory.

When the location of the COP was measured for a number of rackets of different size and shape, it was fairly close to 20 inches from the butt end of all of them (assuming a pivot point 4 inches from the butt end).

If a tennis racket is struck at the center of percussion, the point on the handle of the racket that is the principal pivot (at the base of your index finger) should not move. The series of pictures in Figure 2.3 confirms this. A racket was suspended by the handle approximately 4 inches from the butt end, and the strings were then struck with a ball, first at a position above, next below, and finally at the center-of-percussion sweet spot. The photographs clearly show that when the COP is struck, there is no initial shock at the pivot point (that is, the suspension point).

▶ If you want to minimize the initial shock to your hand, you should try to hit the ball at Sweet Spot 1.

A   *B*   *C*

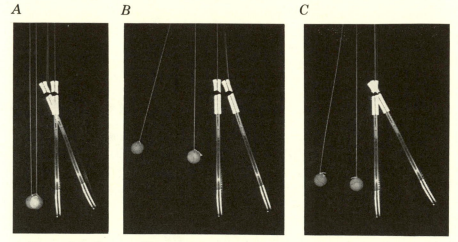

Figure 2.3. Center of Percussion (COP) at Work. A tennis racket is suspended by a thread at the point where the index finger usually grips the handle. The racket is then struck by a tennis ball near the tip (*A*), near the throat (*B*), and at the COP point (*C*) as strobe-flash pictures are taken. In these double exposure pictures, as the ball strikes the strings, you can see what happens to the suspension point (the black dot on the white handle). When the ball strikes near the tip (*A*), the suspension point moves to the right and thus would exert a force on your fingers if you were holding the racket. When the ball strikes near the throat (*B*), the suspension point moves to the left and thus would exert a force on your palm if you were holding the racket. When the ball strikes the COP point (*C*), the racket handle does not move so it exerts no force on either your fingers or your palm.

2.3 Sweet Spot 2: Minimum Uncomfortable Vibration of the Racket and its Effect on Your Hand and Arm (The Node)

When a racket hits the ball, the racket deforms from the impact; then it begins to vibrate or oscillate for a short period of time. The racket can vibrate in several different ways, the most common of which are shown in Figure 2.4. In this diagram, the modes of oscillation A and B can only take place if the racket handle is firmly held in place, while the mode labeled C is for a completely free racket. If a typical tennis racket is vibrating in mode A, it will be oscillating at about 20 to 30 cycles per second. In modes B and C it will oscillate at about 100 to 150 cycles per second. (The stiffer the racket frame, the higher is that racket's frequency of vibration for a given mode of oscillation.) It is the higher-frequency modes of oscillation (B or C) that many people find particularly undesirable. They lead to a loss of control, fatigue, and a generally unsatisfactory feeling when the ball is hit. However, there is one impact point on the racket face (the node) where this vibration is

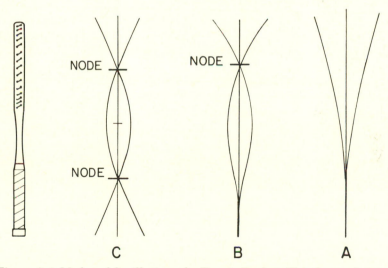

Figure 2.4. Modes of Oscillation of a Tennis Racket. This figure shows schematically how a tennis racket will vibrate. The fundamental mode (A), which has the lowest frequency (with the handle clamped) is sometimes called the diving-board mode, for the obvious reason. Figure 2.2B shows the first harmonic (the next higher frequency) mode of oscillation with the location of the node. When oscillating in this mode, there is no motion of the racket at the node. Figure 2.2C shows the location of the nodes when the racket handle is free rather than clamped.

not produced. This is Sweet Spot 2. The farther from this spot the ball hits, the larger is the amount of this oscillation. The racket will vibrate very badly, for example, if you hit the ball near the tip. The size of Sweet Spot 2 is therefore determined by how much vibration you are willing to tolerate and how quickly the vibrations dampen.

Figure 2.5 shows the oscillation of a racket (with its handle firmly clamped in a vise) as recorded on an oscilloscope when a ball strikes the racket above (A), on (B), and below (C) Sweet Spot 2. When the ball misses the sweet spot, the result is a jagged, harsh oscillation which, if you were holding the racket, you would find unpleasant. When the ball hits the sweet spot, the resulting oscillations are smooth, and you would say the hit feels good or sweet.

Figure 2.6 shows the forces transmitted to the hand from a hand-held racket that is struck above (A), at (B), and below (C) the node or sweet spot. The great reduction in vibration that the hand feels when the ball hits at or close to the sweet spot is obvious from this figure.

Since it is very difficult to hit the ball at exactly this sweet spot every time, many racket manufacturers attempt to reduce (damp) the vibrations as quickly as possible by a variety of techniques. Some use special vibration-damping materials in their frame; others place special material inside the handle; still others put the material between the bridge and the frame; one has a highly damped elastic monomer tuned to the unwanted vibrational frequency inside the racket butt. If these materials do not absorb the energy of the vibration, your hand and arm will. You can buy and install in your racket strings near the throat a small, 1-gram device designed to damp vibrations. So small an object cannot absorb and damp the vibrations of a 350-gram racket frame in a finite time. Instead, these devices very successfully damp the vibrations of the strings, which are oscillating at 500 to 600 cycles per second, and in so doing change the sound of the interaction between ball and racket from a ping to a dull thud. These small objects do not seem to affect how the racket plays in any other way, and since it is not clear whether string vibrations are at all deleterious to your game, using one cannot harm, and may help, your tennis.

Locating the point that minimizes the uncomfortable vibrations to your hand and arm (finding the node of the first harmonic)
Tape a common index card loosely to the handle of your racket (a rubber band can be used instead of tape) and hold the racket be-

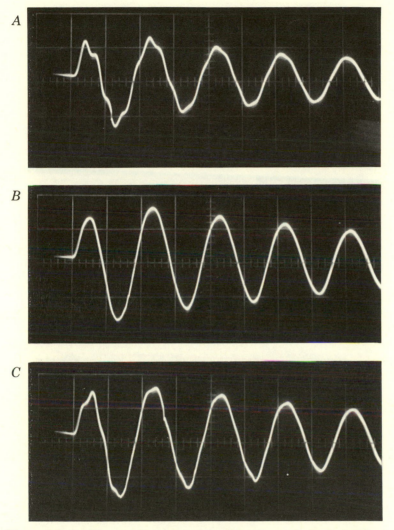

Figure 2.5. Oscilloscope Patterns of the Vibrations of a Racket. When the racket is struck at the node of the first harmonic (*B*), the resulting oscillation does not have a first harmonic (see Figure 2.4), and it is smooth (sweet). The shot feels good. The other two photographs show the oscillation pattern when the racket is struck near the tip (*A*) and near the throat (*C*). The higher frequency vibrations are clearly visible, and a rough, harsh feeling accompanies the shot, just as the pattern is rough or harsh.

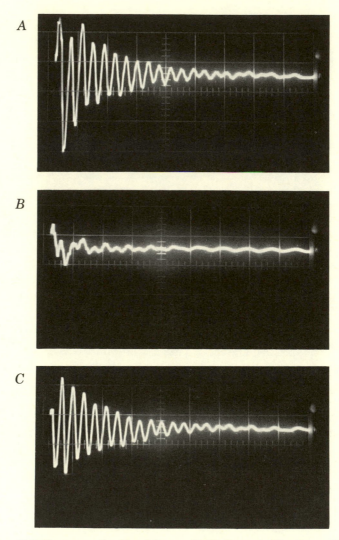

Figure 2.6. Forces on the Hand from a Ball's Impact. The forces that your hand will feel as a ball impacts near the tip (*A*), near the node (*B*), and near the throat (*C*) of a racket are shown. It is clear that the vibration of the racket that occurs when the ball hits near the tip or throat is not present when the ball hits near the node.

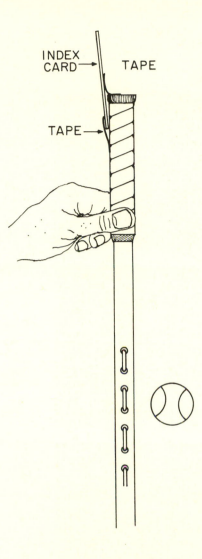

INDEX
CARD→

TAPE

TAPE→

Figure 2.7. Locating the Node. An index card is loosely taped to the handle of the racket and the racket is held loosely about 6 inches from the butt end. When the strings of the racket are then hit with a ball or the butt of another racket, the index card will vibrate and buzz loudly if the hit misses the node (Sweet Spot 2).

tween your thumb and index finger 5 or 6 inches from the butt end, as shown in Figure 2.7. Hit the racket face at various places with a hand-held ball or the butt end of another racket. When you hit the strings near the tip or the throat, the index card will buzz loudly. When you hit at the node, all you will hear is the ping of the strings. As you move the location where you hit away from the node, the amplitude of the buzz will increase. Normally, when you are holding the racket by the grip in the conventional way and you miss the node, your hand and arm absorb all the vibration that the index card indicates is present.

▶ If you want to minimize this uncomfortable vibration of the racket, you should try to hit the ball at Sweet Spot 2.

2.4 Sweet Spot 3: Maximum Rebound Speed of the Ball

On most rackets you will get more power when the ball strikes the strings closer to the throat than when the ball hits in the center of the head. This means that as you move the impact point closer to the throat, you will get more ball speed, at least until you run out of space because the ball hits the frame. A region of even higher power not available in a racket of conventional size and shape becomes accessible on a racket in which the shaft is made shorter and the head made larger by extending the head toward the handle. An enlarged power sweet spot is one of the advantages of rackets such as the Prince, Head Director, durbin, Weed, or the Yonex R7, in which the head has been enlarged in this way.

To take advantage of the power that is available in one of these extended-head rackets, you must learn to hit the ball an inch or two closer to your hand than you would with a conventional racket. If you prefer to use strokes with a great deal of wrist motion, or if you snap or whip your shots, there may be less gain in power overall. For this type of stroke, when you hit the ball closer to your hand, the racket-head speed at the point of impact is lower than the racket-head speed at the conventional impact point. This reduction in speed could more than counteract the higher power that you obtain by hitting the ball closer to the throat of the oversize racket. This is especially true on the serve, where the laws of physics require a high speed of the racket head if you want high ball speed. Because the serve is a shot that allows you to whip or snap the racket (you are actually pronating your arm), the extra speed that the racket has away from the throat can be put to good use and the center of the head may be the best place to get power.

To determine why the point of maximum power is near the throat, not in the center of the head, requires a model of the racket when it hits the ball. Does a real racket act like a racket with its handle clamped in a vise (because your hand is holding the racket there), or does it act like a completely free body because the hand is quite weak compared with the forces involved? Tennis experts and researchers do not agree on the answer to this question, but surprisingly, careful analysis shows that the location of the maximum power point (Sweet Spot 3) is the same in both cases.

If the head of the racket (not the handle) is firmly clamped in place, the point of maximum power will be at the center of the strung area of the head because the strings are "softest" there. If, on the other hand, it is the handle that is firmly clamped (which is a more reasonable way to simulate a hand-held racket), the point of maximum power is closer to the throat. This is so because the racket is flexible, and the energy fed into racket deformation cannot be returned to the ball. The closer to the throat the ball hits, the greater is the effective stiffness of the frame and the less energy is lost in racket deformation.

This may seem at first glance to be wrong, since you were just told that loose strings gave more power (they absorb and then return the ball's energy). Why, therefore, shouldn't a more flexible racket give more power? The reason has to do with the timing. A ball spends about 5 milliseconds (thousandths of a second) in contact with the strings before it is propelled away by the strings snapping back. The racket, on the other hand, takes about 15 milliseconds to return to its undistorted position. By that time the ball is long gone and cannot benefit from any energy that the racket absorbed in deformation.

The coefficient of restitution (COR) is defined as the ratio of the rebound speed of the ball to the incident speed of the ball. Maximum power is obtained when the COR is at a maximum. In the laboratory, the COR can be measured at many places on the face of the racket by clamping the racket handle in a vise and firing balls at the face with a ball machine. High-speed motion or stroboscopic pictures allow a measurement of both the incident and rebound speeds of the ball. It is then possible to obtain the COR for every region of the racket head. The racket manufacturer arbitrarily decides that the sweet spot is that region where the COR is greater than 0.5 (or 0.55 or 0.6) and measures how many square inches of the racket's head meet this criteria.

If the racket is treated as if it were a completely free body, the point of maximum power is as close to the balance point (center of gravity) as possible. That is because the closer the ball hits to the balance point, the less energy goes into racket rotational energy. Therefore more energy goes into the ball, and you get more power from the racket. Since the region of the strings closest to the balance point is the area near the throat, that is the region of maximum power.

Locating the point of maximum power (the maximum of the coefficient of restitution)

If you want to determine the location of this sweet spot yourself, clamp the racket handle in a vise with the face up (strings parallel to the ground). If you do not have a vise or a C clamp, put 4 or 5 inches of the handle on the edge of a table and hold the racket firmly in place by pressing down very hard with the palm of your hand. Drop a tennis ball from a height of a foot or more above the strings and measure how high the ball rebounds. Take other measurements while dropping the ball onto the racket face at various places, taking care to release the ball from exactly the same height every time. The point on the racket face that gives the maximum rebound height is the point of maximum coefficient of restitution or Sweet Spot 3 (Figure 2.8).

▶ If you want to maximize the power of your shots, you should try to hit the ball at Sweet Spot 3.

When you try to find the sweet spot by this method, do not be surprised if you discover that the rebound height (hence the COR) is largest near the throat, falls off gradually as the impact point moves away from the handle, but suddenly increases just as the impact point reaches the tip of the racket. At the tip, the racket is hitting the ball twice, although this occurs in a time span so short that you cannot see it happen. Computer modeling and then a series of high-speed stroboscopic photographs confirm this hypothesis. The initial rebound at the tip of the racket is so low that the ball almost stands still. The racket, however, also rebounds, and when it returns 15 thousandths of a second later, it hits the stationary ball a second time. The ball thus acquires some of the energy that is stored in racket deformation and is usually lost.

2.5 Summary

There are three sweet spots on a tennis racket. They are located where

the initial shock or jar to the hand is at a minimum (the center of percussion)
the uncomfortable vibrations of the racket are at a minimum (the node)

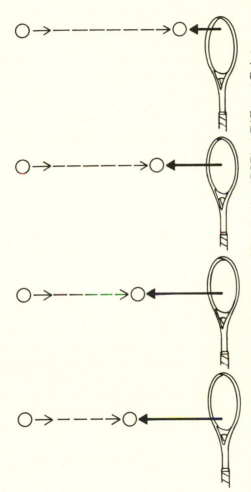

Figure 2.8. Variations in the Coefficient of Restitution (COR) at Different Points of Impact on the Same Racket. The rebound height of a ball for various impact locations on the head of a racket is shown schematically, assuming the ball is always dropped from the same height. The value of the coefficient of restitution (the ratio of rebound ball speed to incident ball speed) is the square root of the rebound height divided by the drop height. If the ball bounces up to 0.49 of its original height, the COR is 0.7 (the square root of 0.49). The larger the COR (the higher the rebound), the livelier the racket will be at that point, since the ball will leave the racket with a higher speed.

the power of the racket is at a maximum (the maximum of
the coefficient of restitution)

The three sweet spots are often found in three different locations
on the face of the racket.

Chapter 3
The Size, Shape, and Weight of the Racket

Rackets come in an assortment of sizes, shapes, and weights; with each of them, the manufacturer is attempting to accomplish something. A manufacturer seldom, if ever, reveals why the head is squared or the throatpiece is inverted or the head is enlarged in some particular way, however, so you have no way of selecting one model over another on the basis of what performance you desire or what your own style of play dictates.

3.1 Width

Although the idea of enlarging the width of the head has been around for many years, and a patent on such a racket was granted in 1974, the first successfully marketed wide-headed racket was Howard Head's Prince in the late 1970s. (By then Howard Head had sold the Head Company and had acquired Prince.) It was known that by increasing the width, the "polar moment of inertia" would also increase. This is the first major advantage of the extra width.

When you hit the ball off-center (not along the long axis of the racket), the racket will tend to twist in your hand and the shot will probably go awry. The property of an object, such as a tennis racket, to resist twisting is called the polar or roll moment of inertia. The moment of inertia is defined as the mass of the object times the distance of that mass from the axis squared. If this moment of inertia is made larger, the racket will be less likely to twist in your hand and will therefore gain stability about its long axis. The moment can be increased either by increasing the mass at the edges of the frame or by making the frame wider. Increasing the width is much more effective because the added mass only increases the moment linearly, while the moment increases as the square of the width. This is why the oversize Prince racket was originally developed and is one of the reasons for its present popularity. A typical oversize racket is about 25 percent wider than a standard size racket, but it has a polar moment of inertia that is 50 percent greater. The Weed racket, which is a super oversize racket and close to the maximum allowable width under the new rules of tennis, is 40 percent wider than a standard size racket and it has a polar moment that is almost twice as large as the moment of a standard size racket. This means for the same off-center hit, the oversize racket will twist in your hand considerably less than the standard size racket, which makes it more stable. You can also increase the polar moment by adding weights along the outside edge of the frame (instead of widening the frame or changing the shape), as Wilson has done on its perimeter-weighted rackets.

The polar moments of inertia of a number of rackets have been measured and are listed with the mass and width of the rackets in Table 3.1. Figure 3.1 is a plot of the measured moments of inertia of the rackets versus their weight times their head width squared. All the rackets tested seem to fall on the same general curve, which shows you that just measuring the width of the head will give you a good indication of the polar moment of the racket and therefore how stable that racket is relative to other rackets.

Although a larger moment does provide, to some degree, an inertia against twisting, it may make the racket feel more cumbersome and possibly more difficult to maneuver. Again you must weigh the advantages for your game and style of play against the accompanying disadvantages. If you often hit the ball off-axis,

Table 3.1. Polar Moments of Inertia for Several Tennis Rackets

RACKET	WEIGHT ounces	MAXIMUM HEAD WIDTH feet	POLAR MOMENT OF INERTIA pound-ft^2
Head Master	12.8	0.781	.0278
Wilson T-2000	13.2	0.794	.0287
durbin aluminum	14.0	0.817	.0323
Prince Graphite 90	12.6	0.869	.0327
Prince Pro 110	12.6	0.902	.0369
Prince Classic 110	13.1	0.912	.0403
Prince Graphite 110	13.5	0.920	.0420
Prince Graphite 125	12.3	0.988	.0453
Weed aluminum	12.5	1.037	.05324

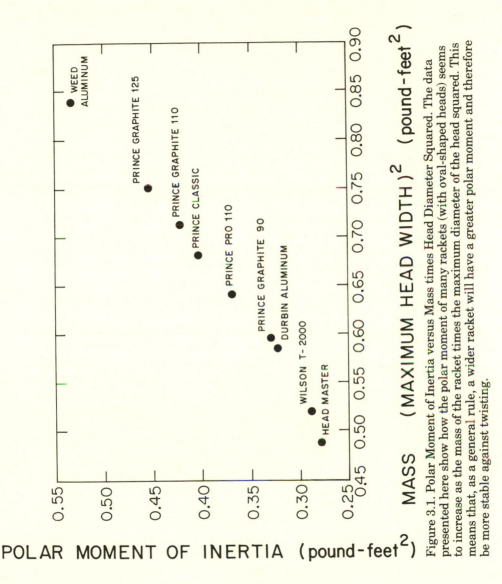

Figure 3.1. Polar Moment of Inertia versus Mass times Head Diameter Squared. The data presented here show how the polar moment of many rackets (with oval-shaped heads) seems to increase as the mass of the racket times the maximum diameter of the head squared. This means that, as a general rule, a wider racket will have a greater polar moment and therefore be more stable against twisting.

causing the racket to twist or turn in your hand, this is one solution you might consider. It is clear that a beginner with this problem would benefit from the greater stability that the increase in polar moment provides.

A point of confusion: A wider head provides more stability when you hit a ball off-center than a standard head does when you miss the axis by the same amount. With the wider racket you can miss-hit by a larger distance; although this will create a greater twisting force (called angular impulse), the same shot might hit the frame of a standard racket. The oversize racket will give more stability than the standard racket, even in this case.

The Spin Shot

A second major advantage of a wider head on a tennis racket is that you will miss-hit shots by striking the frame less often. You do not have to be a physicist to determine why. What is not so obvious is that, because of the wider head, you can hit spin shots with a much larger margin for error and will miss-hit less often. For flat shots, the wider head will increase your margin for error in proportion to the head's increase in size, but you will find it an even greater advantage when you chop, slice, or hit topspin shots. The technical arguments for this extra advantage are as follows.

On a flat, spinless drive, the racket is moving perpendicular to the plane of the strings during the time of impact with the ball (Figure 3.2A). If you aim to hit the ball in the center of the strung area, your margin of error is half the width of the head. In other words, you can miss the center of the racket by almost half a head and still hit the strings rather than the frame. Even if your timing is a bit inaccurate, you will still hit the ball somewhere near the center of the head since your racket is moving along the line of the ball. On a spin shot, the racket is moving at an angle to the plane of the head and at an angle to the direction of the ball's motion (Figure 3.2B). The effective width of the racket head is reduced, and a slight error in timing will cause you to hit the frame and completely bungle the shot. On a flat shot, the ball hits and then leaves the strings from the same spot because the motion of the racket is parallel to the ball direction. On a spin shot, the ball strikes one spot on the strings and then slides or rolls to another spot before leaving the racket (Figure 3.2C). If the distance of this slide or roll is several inches, you will have very little margin of error on a racket of normal width. Using a wider racket does not increase or decrease the distance the ball moves on the strings; it

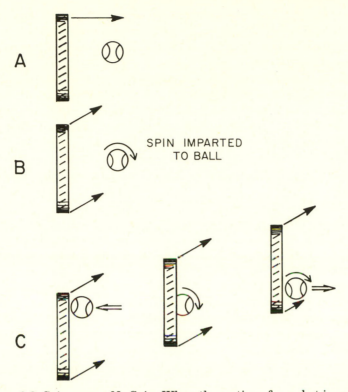

Figure 3.2. Spin versus No Spin. When the motion of a racket is perpendicular to the plane of its strings, a ball that is struck receives no spin; this is called a flat shot (*A*). When a racket moves upward (or downward) so that its motion is not at 90 degrees to the plane of the strings, spin is imparted to the ball (*B*). An upward motion produces forward spin, or topspin, while a downward motion (a chop or a slice) gives the ball backspin, or underspin. On a spin shot, the ball should hit the racket off-center, since it will slide or roll across the strings before leaving the racket (*C*).

just gives you more clearance on each side of it. Bjorn Borg is famous for his topspin shots, and when he miss-hits, the ball usually sails out of the stadium. By increasing the width of your racket, you greatly reduce your chances of doing this, and you will be able to put even more spin on the ball without hitting the frame.

The distance that the ball slides or rolls across the racket face is determined by how you swing and the dwell time of the ball on the strings. If this dwell time is reduced by increasing the string tension, the distance the ball moves across the face of the racket decreases and the margin for error increases. This might be one of the reasons Borg strings his rackets at 75 to 80 pounds, since he would miss-hit fewer balls with his topspin shots.

A third advantage of a wider racket is the way in which the strings play. Synthetic strings seem to play a little "softer," or a bit more like gut in the larger head racket, as was pointed out in Chapter 1.

3.2 Head Length

As was discussed in section 2.4 on Sweet Spot 3, when the head of a racket is lengthened by extending it toward the handle and making the shaft shorter, a region of high power (larger COR) is produced that is not present in a standard size racket. This is because as the point of ball contact moves toward the grip on the handle and the balance point, the racket becomes effectively stiffer. The less the racket flexes, the less energy goes into its deformation and the more energy ends up in the ball, giving you more power.

You do not have to hit the ball close to the throat to appreciate the greater power of the extended head. If you hit the ball in the center of the strung area with one of these oversize rackets, you will get more power than if you hit in the center of the head of a standard size racket. This is because the center of the strings is closer to your hand, and therefore closer to the balance point, on an oversize racket than on a standard size racket.

Extending the head toward the handle will often put the COP sweet spot closer to the center of the strung area of the head, not because the sweet spot will move, but because the center of the head will move toward the handle until it coincides with the sweet spot. For example, on a Prince racket the center of percussion (COP) is within a fraction of an inch of the center of the strung area of the racket head. On a standard size racket the COP is closer to the throat of the racket. This means that when you hit a ball in the center of the strung area of an oversize racket, the initial shock to your hand and arm is less than when you hit the ball in the center of the head of a standard size racket.

3.3 Configuration

When rackets were made primarily of wood, the heads were either oval or circular and had no sharp bends. These were the only shapes strong enough to withstand the forces of string tension and struck balls. And these were the only shapes that could be achieved by steaming, bending, gluing, and clamping strips of wood. With the advent of new metal alloys, strong and light composites, and various space-age materials and techniques, the racket designer was freed from previous limitations. The squared head or inverted throatpiece is used in an attempt to provide more uniform power or COR response over the head of the racket. A circular head with a uniform stringing pattern will have one spot of highest power; the response will fall off as the impact point moves away from this spot. Since it is unlikely that you will hit every ball at exactly the same point on the racket, you will have an unwanted variation in response from shot to shot. If the head shape, string pattern, and string tensions could be modified to reduce the variation in power response at various locations on the strings across the head, you would have a more consistent and predictable shot.

This can be accomplished in many ways, and manufacturers have tried several of them. If the head of the racket is enlarged from 8 to 10 inches in width, for example, a ball missing the sweet spot by 2 inches will be "off" 2 out of 10 inches rather than 2 out of 8 inches. This provides a slightly more uniform response. If the holes for strings are drilled in the frame in a nonuniform pattern so as to give the strings greater density at the COR sweet spot and less density elsewhere, the result will be a more uniform response. If the frame is configured so that many of the strings are of the same length, giving the same tension divided by length, this will lead to a more uniform response. This latter idea can be accomplished by squaring the head somewhat (as Yonex and Snauwaert do), inverting the throatpiece so that its curvature matches the curvature of the tip of the racket (Rossignol, for example) or making the head diamond shaped and stringing it on the diagonal at 45 degrees (MacGregor Bergelin Longstring). This particular shape head allows the racket designer to have the strings enter the frame almost at 90 degrees to the structure. The purpose of this is to have the strings never touch the frame itself, but rest on low friction rollers and act as if they were much longer than the actual size of the head.

Some of this variation of the response of the racket with respect to the position of ball impact is said to be reduced when the string tension is increased. This may reduce overall power, but the increase in control that you acquire may be worth it.

3.4 Weight, Balance, and Moment of Inertia

Rackets come in various weights, from 11 to 14 or 15 ounces. They may be labeled L (light), M (medium), or H (heavy), they may be stamped with their respective weights in grams (100 grams = 3.52 ounces), or they may be completely unlettered and unnumbered. Some manufacturers make only one weight of racket, which they do not trouble to label. Years ago, one usually thought of the big power hitter using a heavy, clublike racket, and the touch or finesse player using a lighter racket. In recent years, there has been a trend away from the use of heavy rackets because of the introduction of new materials, better design, and improved construction. The older style of lightweight frame would either break or become soft in the head from constant hard hitting, so the strong player had to use a heavier racket. In addition, it was very difficult to provide the stiffness that many players desired in the light frames that were available; stiffness required making the frame thicker, which led to heavier rackets. With today's technology it is possible to make a lightweight racket with both stiffness and durability by choosing the right combination of materials and design.

Now that you are no longer constrained in the selection of weight by the old technology, what kind of racket do you select? Does a heavy racket actually give more power than a similar but lighter one? This is a complicated problem, which is best analyzed by making some simplifying assumptions. In this analysis, the racket is treated as a free body and the weight, but not the weight distribution or the balance, is allowed to vary. Using these assumptions, the kinematic relationships between tennis balls and rackets were calculated; the results are shown in Figures 3.3 through 3.5. Figure 3.3 shows that the speed of the tennis ball changes only slightly as you change the weight of the racket, provided you keep all the other parameters (racket speed, incident ball speed, and so on) fixed. This is because a 12-ounce racket is already six times as heavy as the 2-ounce ball; adding another 2 ounces to the racket does not change the ball speed appreciably. As you can see from Figure 3.4, this is also true for a serve, where the initial ball speed is zero. On the other hand, Figure 3.5 shows that when the racket weight is kept constant and the racket-head speed is allowed to vary, the speed acquired by the struck ball depends very strongly on the speed of the racket head. (A rough rule of thumb is that the ball speed leaving the racket is ½ the incident ball speed plus ³⁄₂ the racket head speed.)

▶ The extra racket-head velocity that you can get with a
light racket may more than compensate for its lighter
weight and give you higher ball speeds.

Because the lighter racket is more maneuverable and less

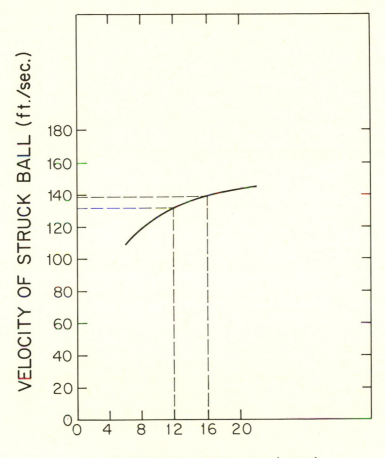

RACKET WEIGHT (OZ.)

Figure 3.3 Ball Speed versus Racket Weight (Groundstroke). This figure
shows what happens to the speed of the ball leaving the racket as you
increase the racket's weight. For a racket-head speed of 80 feet/second,
an increase of racket weight from 12 to 16 ounces only increases the
ball speed from 131 to 138 feet/second. An increase of 33 percent in
racket weight leads to an increase in ball speed of only 6 percent.
(These curves are for a groundstroke with the ball coming at you at
60 feet/second.)

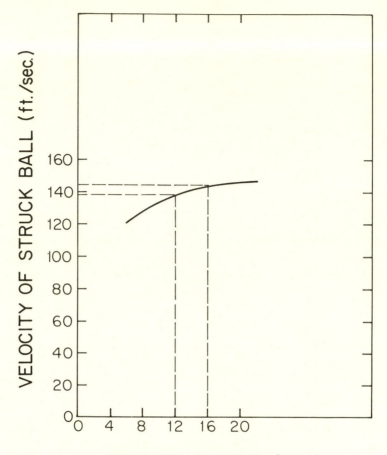

Figure 3.4. Ball Speed versus Racket Weight (Serve). This figure is similar to Figure 3.3, only it is for the serve and not a groundstroke. For a racket speed of 100 feet/second, a 33 percent increase in racket weight (from 12 to 16 ounces) increases the ball speed from 138 to 144 feet/second (or less than 5 percent).

fatiguing to play with, it could give you an additional fraction of a second in which to react to an incoming ball. Although the heavier racket requires more effort to get the same head speed, there is less reaction back on your hand and arm when you hit the ball than there is with a lighter racket. In addition, you will get somewhat better control with the heavier racket on miss-hits because the racket is less likely to twist or turn in your hand. Also, since a heavier racket requires slightly less head speed to obtain a given

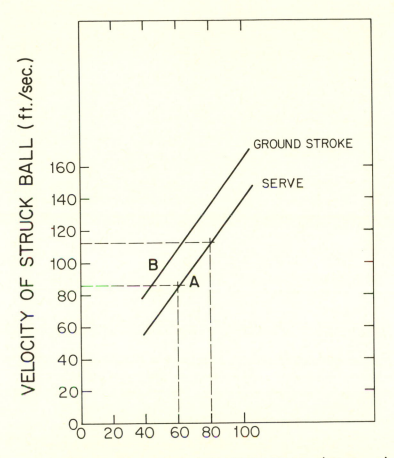

Figure 3.5. Ball Speed versus Racket-Head Speed. This diagram shows how the speed of the ball as it leaves the racket varies when the racket-head speed is changed. The steep slope of the two curves means that the resultant ball speed is highly dependent on the racket-head speed for both the serve and groundstrokes. Increasing the racket-head speed from 60 to 80 feet/second increases the ball speed on the serve from 86 to 113 feet/second (A). To get the same ball speed on a serve as on a groundstroke (100 feet/second), the racket must be moving about 20 feet/second faster (B). Since the ball speed in a serve is usually considerably higher than the ball speed in a groundstroke, the racket-head speed must be *very much* higher when serving—possibly almost twice the racket-head speed of a typical groundstroke.

ball speed, you should have slightly better control of your shots when you swing more slowly, for angular error as the ball rebounds off the racket is proportional to the racket-head speed. (A fuller explanation of this is given in a later chapter.)

Balance and Moment

Total racket weight is not the only consideration, since there are two other parameters that influence your swing, the power you get, and how the racket plays or feels. One is the balance and the other is the *swing weight* or moment of inertia. Is the racket heavy or light in the head, or is it evenly balanced? By convention in the tennis industry, a racket that has its balance point (center of gravity) at its geometrical center is said to be *balanced*. If the balance point is on the head side of the racket's center, the racket is said to be *head heavy*. If the balance point is on the handle side of the racket's center, the racket is said to be *head light*. It is quite easy to find the balance point of a racket, and this location and the racket weight allow you to find the *first moment* of the racket.

When you pick up a racket by the grip and hold it in your hand, you feel a certain weight. This is not the actual weight of the racket (the value that you would get if you put the racket on a scale and weighed it) but its first moment, which is the weight of the racket multiplied by the distance from the balance point (also known as the center of gravity) to the location of your hand. For a typical racket, 350 grams (12.3 ounces) and 27 inches long, with its balance point at the center (13.5 inches from the butt end), this first moment is 117 ounce-inches (12.3 ounces times 9.5 inches, since the hand holds the racket about 4 inches from the butt end and 13.5 − 4 is 9.5 inches). If the racket balanced ½ inch head light, the first moment would be 111 ounce-inches or about 5 percent lighter. If the racket balanced ½ inch head heavy, the first moment would be 123 ounce-inches, or about 5 percent heavier.

You can change the balance of a racket by adding weight (lead tape for example) to the tip of the racket; this will of course change the weight as well. Table 3.2 shows how much the balance point of a racket shifts and the first moment increases as weight is added at the tip. If weight is not added to the tip but to the outside edge of the head (at the point where the racket is widest) it will not be as effective in moving the balance point and increasing the first moment. Weight added at this location will increase

Table 3.2. Location of Balance Point as Weight is Added to Racket Tip

Grams[a] ADDED	Location of BALANCE POINT *inches*	Shift of BALANCE POINT *inches*	First moment FROM HAND *ounce-inches*
0.00	13.50	0.00	117.08
1.00	13.54	0.04	117.89
2.00	13.58	0.08	118.70
3.00	13.61	0.11	119.51
4.00	13.65	0.15	120.32
5.00	13.69	0.19	121.13
6.00	13.73	0.23	121.94
7.00	13.76	0.26	122.75
8.00	13.80	0.30	123.56
9.00	13.84	0.34	124.37
10.00	13.87	0.37	125.18
11.00	13.91	0.41	125.99
12.00	13.95	0.45	126.80
13.00	13.98	0.48	127.61
14.00	14.02	0.52	128.42
15.00	14.05	0.55	129.23
16.00	14.09	0.59	130.04
17.00	14.13	0.63	130.84
18.00	14.16	0.66	131.65
19.00	14.20	0.70	132.46
20.00	14.23	0.73	133.27
21.00	14.26	0.76	134.08
22.00	14.30	0.80	134.89
23.00	14.33	0.83	135.70
24.00	14.37	0.87	136.51
25.00	14.40	0.90	137.32
26.00	14.43	0.93	138.13
27.00	14.47	0.97	138.94
28.00	14.50	1.00	139.75
29.00	14.53	1.03	140.56
30.00	14.57	1.07	141.37

NOTE: Figures are based on a 350-gram (12.3-ounce) racket with its balance point 13.5 inches from the butt.

[a] one ounce = 28.4 grams.

the polar moment of inertia of the racket, however, which will increase the racket's stability against twisting on off-center hits.

The other parameter, swing weight or moment of inertia about an axis perpendicular to the racket handle (sometimes called the second moment), may also be important to you as a player. This is the same quantity that was discussed with regard to stability against twisting, but here it is measured about a different axis in units of mass times length squared. Unlike the first moment, which tells you how heavy the racket feels when you hold it in your hand, the second moment tells you how heavy the racket feels when you swing it. This moment of inertia is hard to define in nontechnical terms other than to say it is the property of the racket that resists rotation. Unlike weight or balance point, it is difficult to measure. These are probably the reasons why no manufacturer supplies the value of the moment of inertia of its rackets. Since the racket is not whipped about on most strokes, the swing weight is of major importance only for the serve.

If you are big and strong, and a tennis racket feels like a table tennis racket to you, then the heavier, head-heavy, large-moment-of-inertia racket is just what you need. For the rest of us, a lighter, lower-moment racket may fit our needs better. This is especially true when serving, since computer studies show that racket-head speed is extremely important in that stroke because the initial ball speed is nil. The only way to give the struck ball very high speed on a serve is to have a very high racket-head speed. To do that, you must be able to whip the frame around hard. On groundstrokes, high racket-head speed is not as necessary, since you can get high ball speed by feeding off the energy that the ball brings in. Also, the typical ball speed produced by a groundstroke is less than the ball speed desired by most servers.

▶ A racket that is light and head light will usually be a good serving racket.

You clearly should match the weight and balance (or moment) of your racket to your style of play. The lighter, head-light, low-moment racket is not only good for the big server, it also is more maneuverable, which is essential for play near the net. This is the racket for the serve and volley player. The groundstroke player who plays on slow courts can use a heavier, larger-moment racket effectively. The all-around player should use a medium-light racket, especially if playing on fast courts.

There is one more argument for a head-light racket that is so obvious you never see it in print. When you buy a head-light racket and then find that you need more weight, you can easily add lead tape to obtain whatever weight, balance, or moment you want. If you buy a heavy or head-heavy racket, there is not much you can do to correct it if you decide you do not like the balance or feel of it.

3.5 Racket Flex or Stiffness

If you hold or clamp the handle of a racket and push on the tip of the head, you will observe a deflection. A racket that deflects or bends a great deal is said to be flexible; if there is very little displacement of the tip, the racket is said to be stiff. You can get a rough measure of the flexibility of a racket yourself (or you can rely on the manufacturer's specifications)—but having done so, you must decide whether you want a stiff or flexible racket. When you hit the ball, how does the bending of the frame influence your shot?

Flexibility and Shot Power

Many experts and some articles and books claim that a flexible racket provides more power. Insofar as we know, however, there are no data to support this claim. The statement that "flex means power" is as erroneous and out-of-date as the claim that "tighter strings give more power," which one finds in old books and articles on tennis but only occasionally in recent publications. A flexible, whiplike shaft on a golf club may produce a drive with more distance, but it is stiffness, and particularly stiffness in the head, that gives a tennis racket its power. In addition, a stiff racket allows a player more control.

A Case for the Degradation of Power

Are there any advantages to a flexible racket? A more flexible racket is certainly kinder to the arm, just as more flexible strings are. The flex of the racket will take up some of the shock and spread it over a longer time. If you have a weak hand grip or arm, or if you have a sore arm or tennis elbow, then a very stiff racket may not be for you. If you are not particularly skilled and tend to hit the ball all over the face of the racket, the flexible racket has its advantages, for the added flexibility on off-center shots will be kinder to your arm. A racket with a certain degree of flexibility is said to be *forgiving* in that it allows a player to make small errors in where the ball is struck on the strings, and it does not punish the arm and elbow as much.

A second way to reduce the jar or shock to your arm is to use a stiff racket that is strung at a lower tension. This has the advantage of not degrading the power, but it may decrease the control you have over your shots.

The stiff racket will give more control, and for the better player, it is strongly recommended. The twisting and bending of the frame of a flexible racket may cause off-center hits to leave the face at unanticipated angles, which prevents you from placing the ball where you want it to go. In addition, when the ball strikes the face of a flexible racket at different spots, the resulting variation in power may also lead to control difficulties.

A stiffer racket will give you both more power and more control, so racket manufacturers try to produce less flexible frames for their top-of-the-line models. Be cautious, though, about buying a very stiff racket and having it strung at very high tension. You will have good control, but you may pay for it with a sore arm or elbow. One possible compromise is a racket with a very stiff head and some degree of flex in the shaft, since flex in this location does not degrade the racket's power as much as flex in the head.

3.6 Material and Composition

At one time, not long ago, all rackets were made of wood. Then metal (steel and aluminum) rackets were developed and successfully marketed. Now the composite-material racket is available as well. Each advance in racket material has allowed the racket designer to do things that previously were impossible owing to the structural limitations of materials.

Woods

Laminated wooden rackets, the old standby for centuries, were long the standard against which all other rackets were compared for power, control, touch, and so on. But wooden rackets have changed a good deal. Today there are not many available on the market, and most are in effect composite rackets because various other materials are used to enhance the properties of the wood. Because it is a cellular material, wood lacks the strength-to-weight ratio of modern composites. Consequently, to get the stiffness or the strength players want, frames made entirely of wood have to be made thicker; they are therefore heavier. Now, by adding material such as fiberglass or graphite, it is possible to make a woodlike racket that is relatively light and has the desired strength and stiffness.

If you give a wooden racket a pounding by playing a hard-hitting game, or if you have it restrung often, the head will soften, and so will your game. If you do not hit the ball hard, and a set of strings lasts the life of the racket, then this will not bother you. Otherwise, the racket's properties will change as you play, and you will have to replace the racket on a regular basis. Because of the limited strength of wood, it is difficult to build a wooden racket that is oversize or has sharp bends in the frame. Oversize rackets, which should be strung at proportionally higher tensions, must be made with considerably stronger frames to withstand the greater forces—even before a ball is hit.

Some players claim that a wooden racket has a certain feeling of liveliness, which a metal racket entirely lacks and a composite racket does not quite have. This is a difficult claim to test in the laboratory. There is no doubt, however, that the composite rackets available today are superior in playing characteristics to the best of the all-wood frames. They should last much longer as well.

Metals

Although steel rackets were initially popular, almost all of the high-quality metal tennis rackets available today are an alloy of aluminum. Possible new challengers are magnesium and titanium rackets, some of which are becoming available. The strength and stiffness of metal frames are controlled by the alloy used and the cross-sectional shape of the aluminum beam that forms the frame. A major improvement in metal rackets is the use of a plastic throatpiece instead of a throatpiece of metal, which was riveted or welded in place and tended to break from fatigue in early metal rackets. Aluminum rackets are strong, can be fairly light, are easy to fabricate (which makes them cheaper), and will last a reasonable time before metal fatigue sets in and requires you to purchase another racket. They do not damp vibrations as quickly as some people might desire; in lab tests they tend to "ring" a bit, rather than absorb the oscillations internally.

Because metal rackets have a thinner frame, you might expect to profit from lower air resistance. At the relatively slow speeds at which a racket head moves on the court, however, this is not a major advantage. The beautiful aerodynamic appearance of some rackets is just that—appearance.

Composite Materials

There are many types of composite rackets available today, and many of them employ the latest space-age technology. One type of composite consists of fibers or filaments of a very strong material (graphite, glass, boron, ceramic) embedded in an epoxy or resin that is then molded into the shape of a tennis racket. Another type of composite racket is a laminate of various materials such as wood, glass, aluminum, and so forth. The purpose of mixing these materials is to obtain a racket with certain flexibility, structural strength, weight, and other properties that no single material possesses. In addition, since these reinforcing materials are fibrous, it is possible to orient the direction of the fibers to give strength and stiffness in one direction and allow some flexibility in another. Happily, many of these efforts have been very successful. Because the strength-to-weight ratio is no longer the major concern in working with these materials, manufacturers can produce oversize frames that weigh under 12 ounces, frames that are very thin, heads that have sharp bends in them, and so on. In addition, these rackets, when fabricated properly, excel in their longevity of use and uniformity. The playing characteristics of a good composite racket, for example, should not change as you use it over the years, and there is no reliable estimate of the lifetime of a good graphite racket. (The same cannot be said for a wooden racket or the strings that you put into the graphite racket.) Another advantage of composite rackets (and metal rackets) is that all the rackets of a certain make are essentially identical, which is not true of wooden rackets, regardless of how carefully they are made.

If you were to saw one type of graphite racket in half, you would find that the graphite is only a thin skin—about as thick as heavy cardboard. Because composite materials are incredibly strong, only a thin layer is needed to give the racket its required strength. If, as you play, you tend to scrape the top of your racket on the court, the tip may begin to show signs of wear from the frequent abrasion or even wear through, in which case you will need a new racket. Protect the racket tip by using a guard or covering it with a tape that most pro shops have just for this purpose. It is silly to ruin a two-hundred-dollar racket for lack of a one-dollar piece of tape.

3.7 Grip and Handle

Tennis experts disagree among themselves about how firmly a player should grip the racket handle. Several people have done experiments, but they have come to opposite conclusions. Some data show that if you grip the racket very tightly, you will have more powerful shots. One expert claims that grip tightness is the most important factor in determining the power you get from your shots. Other data show that the ball speed and racket recoil are independent of whether you clamp the handle firmly in a vise or allow the racket to be completely free of all restraints. From what we have learned about the COP sweet spot, we can see that if the ball hits exactly that point, the importance of grip firmness will be reduced. If the ball hits any other point on the face of the racket, it will try either to twist or turn the racket in your hand or to break the racket from your grip on the handle. How can these seemingly contradictory results be understood?

Let us examine the physics of the situation. Theoretically, if a racket were perfectly rigid, then when its handle was firmly held in a vise, the maximum coefficient of restitution would be about 0.9 in the center of the head. When a real, and therefore flexible, racket is tested in a vise, the COR varies from 0.6 near the throat to 0.2 near the tip. The actual values depend upon the racket's longitudinal and transverse stiffness and its inertial properties. A more flexible racket will have lower values of COR than a stiff racket as the ball's impact point approaches the edge of the frame.

What about the "free" racket? The COR of a free racket, calculated assuming it had its measured inertial properties but was absolutely rigid, and the COR of the same racket tested with its handle held in a vise were similar. Were flex to be added to the free-racket calculation, the results would show that a free racket does not have the power of a clamped racket.

It is not possible, then, for a free racket to give as much power as a clamped racket. Thus, the main advantage of gripping the racket handle tightly is that you will get a little more power. If you hit the ball at, or very close to, the center of percussion, the tightness of your grip is not particularly important. If you miss the center of percussion, the COR will be a bit lower if you are not holding on very tightly. The other principle advantage of a reasonably firm grip is to keep the racket from slipping, twisting, or coming out of your hand when you do not hit a shot exactly where you want to on the strings.

Grip Material

Grip material can wear out. Most rackets come with a leather grip, which you should replace or recover quite often. Since the grip is your contact with the racket, and since a new grip is not expensive when compared to the racket or strings, there is no excuse for not replacing the grip or recovering the handle. There are a number of materials (such as a gauze) that you can simply wrap over your present grip and then strip off when it no longer works. Some of the top professional players do this as they change ends between games. There are all types of synthetic materials that are quite porous and can absorb a good deal of moisture. Some materials provide better friction when wet than dry, or so their manufacturers claim. These synthetics, as well as leather, are readily available, and none of them takes more than a few minutes to install on your racket. There are also cleaners available that will make your present leather grip less slippery and will extend its life. Regardless of how you maintain the grip material on your racket, proper care should save you from one of the worst feelings you can experience on court: to hit the ball off-center and feel the racket slipping.

Grip Size and Handle Shape

The handles of tennis rackets are eight sided and come in several girths or sizes, the smallest being about 4⅛ inches in circumference, and the largest 4⅞ inches. The steps from the minimum to the maximum size are uniform increments of ⅛ inch. The size is often given on the racket handle as a number from 1 to 7, where that number is the number of eighths of an inch that the circumference exceeds 4 inches. Therefore, a racket with a 4⅝-inch handle will be labeled with a 5. It is not uncommon for the size shown on the racket to be off by ⅛ of an inch when you actually measure the girth of the handle.

Manufacturers configure their racket grips in a variety of octagonal shapes; some are nearly square, others are so flattened as to be almost round. When buying a racket, most players seem to choose a grip size that feels comfortable but they do not consider the shape of the handle. You should test the various shapes and see if one feels more comfortable than the others. The average tennis player knows from experience whether he is holding the racket with a forehand or backhand grip without actually looking, but for a beginner this is not obvious. A grip that facilitates one racket position over the other can be used to great advantage.

Using Torque to Prevent Twisting

The physical quantity that you apply to prevent the handle from turning or twisting in your hand is called torque; it is the product of the frictional force between your hand and the grip material, and the average radius of the handle at the grip. The frictional force is the product of something called the coefficient of friction (COF, a measure of how rough or slippery something is) and the force with which your hand presses down on the handle. As an equation, this would be:

$$torque = COF \times force \times radius.$$

The COF goes down as the grip gets slippery, and you know from experience that the racket is more likely to twist in your hand (there is a lower torque) when the handle is slippery. The tighter you grip the handle (higher force), the less likely the racket is to slip (there is a higher torque). Hence, to reduce the chances of the racket twisting in your hand, you should use a good, high-friction, nonslippery grip covering, squeeze the handle firmly, and use the largest size grip that is comfortable for you. This last recommendation is not as obvious as the other two are, but it follows directly from the physics, and it is important. The radius referred to in the above equation is the handle size (in inches) divided by 2π, which is about 6.28. An increase of an eighth of an inch in grip size will increase the available torque by only a few percent, but why not take every advantage that you can? If this were the only consideration, you would use the largest possible grip instead of the one that feels most comfortable, but there are two other things to remember about large grips. First, if you use too large a grip, your hand may tire and the torque you can apply may be diminished the longer you play. Second, a larger grip will tend to lock your wrist on your swing, while a smaller grip will allow you to use more wrist in your shots. This is something to consider when you are choosing the size of your grip, because you may or may not want to have a firm wrist depending on your style of play.

3.8 Summary

A wider racket head will provide more stability, and a greater margin for error, particularly on spin shots.

A racket head extended toward the handle will have a region of higher power and will have the COP closer to the center of the head.

A stiff racket will give you more power and control.

Racket weight and balance should be matched to the style of play. Light, head-light, low-moment rackets are best for the serve and volley game. Medium-weight, balanced rackets are best for the player who stays in the backcourt and hits groundstrokes.

Chapter 4
Understanding the Motion of the Ball
(The Bounce)

The motion or path or trajectory of a tennis ball through the air is completely determined by the laws of physics. The height to which the ball bounces and the "speed" of the court are also subject to the same laws. But you do not need to be a physicist in order to use physics to play better tennis, for you have learned from experience to take the first few feet of a ball's trajectory and anticipate exactly where that ball will be at some later time. From years of watching tennis balls, you know instinctively to start to move in or back up, how high the ball will bounce, or how fast it is coming at you. You do not have to think about it—and you certainly do not calculate the trajectory in your head by applying Newton's laws of motion. You do a pattern recognition to match the pattern of a ball's trajectory.

Knowing the physical laws behind the trajectory and the bounce of a tennis ball, however, can be of use in a number of situations and can make you a better player. You will not need to apply these laws yourself, but by using the results of the analyses provided in these next chapters, you will be able to master and understand what is happening to the ball 99 percent of the time. (The ball spends less than one percent of the time in contact with a racket and 99 percent of the time traveling between hits.)

4.1 The Bounce of the Ball (without Initial Spin)

Tennis courts are made of all types of surfaces—clay, grass, wood, concrete, asphalt, plastic, canvas, composition, and rubber, to mention just a few. There are very stringent rules on the length and width of a court, but there are none concerning its surface. It is assumed that the court is level and smooth, but even that is not specified in the official rules of tennis. Some types of courts are said to be *slow*, while others are described as *fast*. What is it that gives a court surface its playing characteristics and makes it either fast or slow?

When a ball bounces on the court, its horizontal speed is usually reduced somewhat by its interaction with the court surface. If the ball slows down a great deal upon bouncing, the court is slow, while a fast court does not affect the ball's horizontal speed as much. There are only two characteristics of a court surface that influence what the ball does when it bounces. These are the coefficient of restitution (COR) and the coefficient of friction (COF) between the ball and the surface. The COR tells you how high the ball will bounce if you drop it from a given height. A high-COR court surface gives a higher bounce than a low-COR surface. The COR is defined as the ratio of the vertical ball speed after the bounce to the vertical ball speed before the bounce; this is essentially the same definition that was used in the second chapter when referring to the ball hitting the racket. The coefficient of friction (COF) is a measure of the frictional force of the court surface on the ball, in a direction parallel to the surface; it usually slows the ball down. A high value of the COF means that the frictional force on the ball is large.

Court Speed and Friction

When the bounce of a ball is analyzed, it is clear that friction is much more important than the COR in determining the speed of the court. While the COR influences the vertical velocity of the ball, the friction affects the horizontal velocity of the ball, and that is the direction that determines a court's speed. The larger the friction between the ball and the court, the more the ball will slow down when it bounces, and the slower the court will be. When a ball with no spin hits a court surface, there is a frictional force parallel to the surface and in a direction opposite to the ball's direction of motion, as is shown in Figure 4.1. The ball will begin to slide or skid along the court, with the bottom of the ball slowing down more than the rest of the ball; this will cause the ball to begin to rotate. If the frictional force is great enough and the ball's incident angle of bounce is large enough, the ball will begin to roll on the court surface before it rebounds and loses contact with the ground. If the ball leaves the court before rolling begins, it is considered to be a fast court. Once this rolling begins, the frictional force is greatly reduced, and the ball does not slow

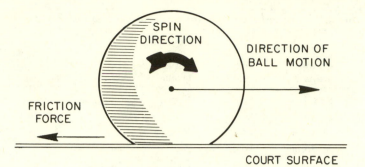

Figure 4.1. Friction Force on a Ball. This figure shows the force of friction on a ball when it bounces and the direction of the spin that this force causes the ball to acquire.

down any more. The exact conditions for loss of speed by a ball bouncing are determined by the angle at which the ball hits the ground and the coefficient of friction. For low friction and a small incident angle, the ball will not begin to roll and will lose only a fraction of its forward speed. For high friction and a large angle of incidence, the ball will begin to roll and will lose 40 percent of its forward speed. Regardless of the friction or ball angle, this 40 percent is the maximum loss of speed that a ball can have in the forward direction when it bounces, assuming it has no initial spin. In other words, the laws of physics determine how slow the slowest court can be. For certain types of shots, such as the lob, there is no difference between a fast court and a slow court because the angle of incidence is so large that the ball will begin to roll regardless of the coefficient of friction. For shots with a small angle of incidence (a low drive, for example), the ball will not roll, so the speed loss will be proportional to the friction between the ball and the court. Table 4.1 lists these various possibilities, and Figure 4.2 illustrates the effects of friction on loss of ball speed graphically.

Many hard courts (Laykold, for example) must be resurfaced quite often if the slowness that they have when they are new is to be retained. These courts are covered with a latex or acrylic paint containing sand. The roughness of the sand creates a great deal of friction between the surface and the ball. As the court is played on, however, constant wear tends to smooth the surface, reducing the friction. As a result, the court speeds up as it ages and is used. Measurements on a newly surfaced court and on the same court many months later show a clear deterioration in the COF. This is predominantly due to wear by tennis shoes, not weather, because

Table 4.1. Incident Angle, Friction, and Ball Speed Loss

INCIDENT ANGLE	COEFFI- CIENT OF FRICTION	CORRES- PONDING COURT SPEED	LOSS OF FORWARD BALL VELOCITY
Small	Low	Fast	Small, but proportional to
Small	High	Slow	friction
Medium	Low	Fast	Medium
Medium	High	Slow	Large
Large	Low	Fast	Maximum
Large	High	Slow	Maximum

areas that had little traffic, such as the alley near the net, show little sign of deterioration. Many hard courts are not resurfaced until there are obvious signs of wear, such as cracks and pits in the surface. This is wrong. A court should be resurfaced when the playing characteristics are no longer the ones you want, even if the surface still looks beautiful.

The Angle of the Bounce (with No Initial Spin)

In addition to slowing down the ball's horizontal speed, friction causes the angle at which the ball rebounds to change. The more the friction, the larger the rebound angle will be; the less the friction, the smaller the rebound angle. On a fast surface the ball seems to skid or slide, hence it does not seem to rise as high, because the rebound angle is smaller. The height to which the ball rises is actually the same for fast and slow courts having the same COR. It is only the angle of rebound that changes. Since most courts do have similar COR values, players tend to judge the speed of the court by noting the rebound angle. The eye and brain are much better at gauging an angle than observing a change in ball speed after the bounce. If the ball comes at you at a high angle after the bounce, experience tells you that the court is probably slow and the ball will be delayed in getting to you. If the ball comes off the court at a low angle after the bounce, you know that the court is fast and you should get your racket ready a little quicker because the ball will be coming at you faster. If you have a fair amount of tennis experience, you make this adjustment without thinking about it.

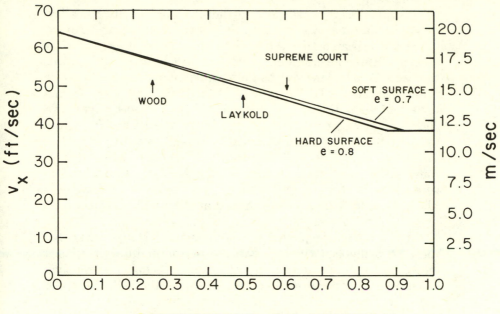

μ, COEFFICIENT OF FRICTION

Figure 4.2. Horizontal Speed of a Ball After the Bounce. The horizontal speed (v_x) of a tennis ball after it bounces is shown for a ball hitting the court with an initial horizontal speed of 64 feet/second and at an angle of 14 degrees. This figure shows that as the coefficient of friction increases, the ball's horizontal speed after the bounce decreases until it reaches a minimum value of 60 percent of its original speed. The coefficient of restitution (e, which tells how high the ball bounces) has very little effect on the ball's horizontal speed as the two curves ($e = 0.7$ and $e = 0.8$) show. When the friction is zero, the ball loses no speed. When the friction is a maximum, the ball is rolling and maintains only 60 percent of its original speed. The three arrows indicate the measured coefficient of friction for a tennis ball on polished wood, Laykold, and Supreme Court. It is clear that the ball has lost more speed on the slow (high-friction Supreme Court) court than on the fast (low-friction wood) court. The tennis shot illustrated here was a drive that hit the court at a low angle; it would roll only if the coefficient of friction were very high (above 0.9).

It is possible to construct a surface that is slow, but has a low rebound angle that would normally indicate a fast court. An example of this is an artificial carpet or surface that is used to cover other surfaces indoors. Measurements on one of these surfaces (Supreme Court) showed it had a high friction and was therefore a slow court. A low COR, however, gave the ball a lower rebound angle and made the court appear faster than it was. Figure 4.3 shows the angle of rebound for two incident angles (26 degrees and 20 degrees) on three different surfaces: wood (COR = 0.8, COF = 0.25), Laykold (COR = 0.8, COF = 0.49), and Supreme Court (COR = 0.7, COF = 0.61). If the Supreme Court surface were as lively as the Laykold (if its COR were 0.8), its angle of rebound would be even larger than that of Laykold because of its large frictional force.

If you have ever played on clay courts with tape lines, you know what happens when the ball hits one of the lines. It skids or slides, does not slow down, and rebounds at a lower angle than expected. This is because the friction between the ball and tape is very low, while the friction between the ball and clay is quite high. The tape acts as if it were a fast court, and the clay is of course slow. You can have a similar, but not so pronounced, experience on a newly resurfaced Laykold court. Balls hitting a white line will not seem to bounce quite the same way as those hitting the colored surface. The friction between the white paint and the ball is somewhat less than the friction between the colored paint and the ball because no sand is put into the white paint. In order to get neat, straight lines, a thin paint is needed; to bind the sand, a thicker paint would have to be used, and this would raise the line higher than the surrounding surface. The variation of bounce on the lines is well known; during the U.S. Open, it was noted on television that balls hitting the lines do not bounce the same as normal shots.

4.2 Adjusting to the Surface

Some people like to play on slow courts, which give them a chance to run down and return shots since the ball really slows up when it bounces. They are usually very steady baseline players, and because they make fewer mistakes than their opponents, they do not want their opponents to win points by hitting shots that they cannot reach. The slower the court, the less likely it is that their op-

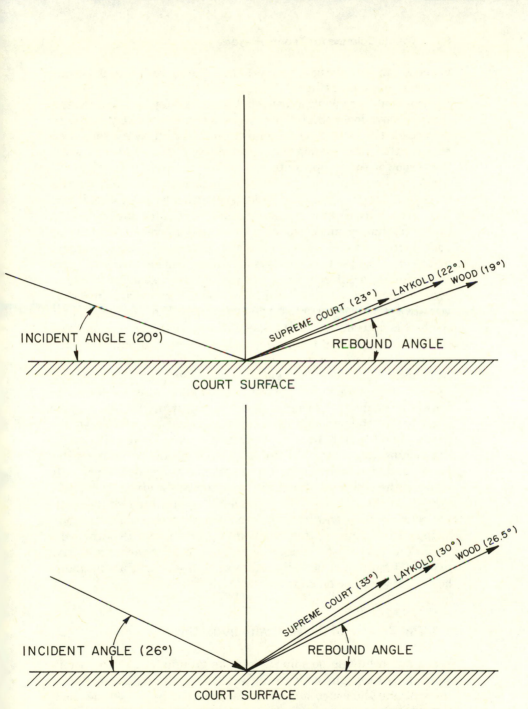

Figure 4.3. Bounce Angles. This figure shows the angle at which a ball rebounds from the ground on three different types of surface when a ball hits the ground (with no initial spin) at 26 and 20 degrees.

ponents can hit outright winners, and the more likely it is that they can win long rallies.

Some players want a fast court because they do not want the court to slow down their shots, allowing their opponents time to get set for the return. These players usually hit every shot very hard, or they have strong serve and volley games, and they want to overpower their opponents.

You should try to play on the surface that is best for the strokes and style of play you have developed over the years. If you have grown accustomed to playing on a particular surface and it is not one that optimizes your game, you might consider changing your style of play. You may try to change strokes, style, strategy, and so on, but that is not always easy. There is one thing you can do to adapt to a different surface if you do not want to modify your style of play. As was discussed in section 1.2 on string tension, if you always swing your racket the same way, the pace you get on the ball is determined by the speed of your opponent's shot since your racket converts the incoming ball energy to outgoing ball energy. Therefore, on a slow court, where your opponent's shots will slow down before they reach you, your shots will lack pace as they leave your racket and will be really slow after they bounce. Rather than changing your style of play by swinging harder, you could reduce the tension in your strings a few pounds; this would give you additional power to compensate for the two effects that are robbing you of power when you play on a slow court. If, on the other hand, you find yourself on a court that is quite fast, you will have all the power you need with your normal swing. Your problem will be to control your shots when they are hurried (because the court is fast) and hard. If you increase the tension in your strings a few pounds, this might give you the extra control you need. You could also change the balance of your racket slightly, making it a bit more head light for use on fast courts or a bit more head heavy for use on slow courts.

4.3 The Bounce of the Ball (with Initial Spin)

When a ball having no spin strikes the ground, it acquires a forward spin or topspin owing to its motion and the friction between the ball and the court surface. This slows the ball down because there has been a force (friction) acting against the direction of motion and some of the energy that the ball had has been converted to spin energy. When a ball having topspin hits the ground, it will not slow down as much as the spinless ball does. Under certain circumstances, it may end up with a greater forward speed than it had before the bounce. To produce this increase in the forward speed of the ball after the bounce, the ball must be spinning so fast that its rotation (in revolutions/second) times its circumference is greater than its linear speed. Most spins and speeds encountered in tennis are not great enough to produce this effect, except possibly a very heavy topspin lob. Hence, a ball with topspin does not jump forward on the bounce; instead, it slows down less than a spinless shot would. Since you base your expectations on the spinless bounce, the ball with topspin bounce seems to shoot forward when it bounces, even though it simply is not slowing down as much as you think it should.

When a ball with backspin strikes the ground, it may act as if it had no spin at all, it may slow up more than a spinless bounce, or under certain circumstances, it may even bounce back somewhat. If the ball trajectory, speed, and court surface are such that a spinless ball would not have begun to roll (it would have skidded or slid), then a ball with backspin will do exactly the same. If the conditions are such that a spinless ball would have begun to roll, a ball with backspin will slow up more than the spinless one (it will "sit up"). What is being said here is this: If you play on fast courts and hit hard, low shots, the ball will skid or slide upon bouncing whether you put backspin on it or not. If you play on slow courts and hit soft shots that clear the net by a good margin, the ball will slow down more upon bouncing when you use backspin than it will when you hit flat shots. Note that these statements are comparing backspin and flat shots having the same trajectory (hitting the ground at the same angle and speed). More will be said about trajectories of spin shots shortly in this chapter and in Chapter 5.

Bounce Height

Tennis players correctly believe that a topspin shot bounces higher than a normal shot. Players anticipate and judge the height to which a ball will rise after the bounce based on their experience with the surface that they are playing on and the maximum height that the ball had in its trajectory before it bounced. For a normal spinless shot, the ratio of how high the ball bounces to its previous trajectory peak is almost the square of the coefficient of restitution. (There is a small correction for air resistance.) This is not the case when the ball has spin. Topspin makes the ball dive, hitting the ground at a higher speed than it would if it had no spin. Since it hits with greater speed, it rebounds with greater speed and thus bounces higher. A ball with backspin floats gently to the ground with a lower speed than the equivalent spinless shot for the same maximum trajectory height. Hence it rebounds to a lower height. To quantify all of this, a computer was programmed to analyze balls being hit waist high, with a speed of 67 miles an hour and at a variety of angles with topspin, backspin, and no spin. The maximum height of the ball trajectory was 58 inches (above the ground) in each case. The results are given below and in Table 4.2.

High, Low, and In-Between Bounces

A ball dropped in a vacuum from a height of 58 inches will hit the ground with a speed of slightly over 17 feet/second. A tennis ball in air, with no spin and with a peak in its trajectory of 58 inches, will hit the ground with a vertical velocity component of 16.3 feet/second (the difference is due to air resistance). If the COR is 0.75, the ball will rebound to a height of 28 inches. A ball with a backspin of 32 revolutions/second and having the same peak in its trajectory will float to earth, hitting with a vertical speed of only 14 feet/second. It will rebound to a height of only 20.5 inches. A shot with a topspin of 32 revolutions/second and having the same 58-inch peak in its trajectory will dive toward the ground, hitting at 18.1 feet/second. It will rebound to a height of 34.4 inches. Clearly, topspin shots do bounce higher.

There is one note of caution about these points. Some people confuse the angle at which a ball rebounds from the court with

Table 4.2. Bounce Height and Ball Spin

Maximum height before bounce inches	Spin rev/sec	Vertical velocity when hits ground feet/sec	Maximum height after bounce inches
58	32 (back)	14.0	20.5
58	none	16.3	27.9
58	32 (top)	18.1	34.4

the height the ball eventually reaches. The rebound height is determined by the ball's vertical velocity when it hits the ground and the COR of the court surface (which gives the vertical velocity as the ball leaves the surface). The angle at which the ball rebounds is determined by the ratio of its vertical velocity to its horizontal velocity immediately after the bounce. It is possible that a ball with a small angle of bounce will eventually reach a large height—if the ball is allowed to continue until it hits the fence.

4.4 Summary

The friction between the ball and the court determines the speed of the court. Fast courts have low friction (they are slippery); slow courts grab the ball.

Topspin shots bounce higher than flat shots; backspin shots bounce lower than flat shots.

You can adjust to different court surfaces and speeds by modifying your game, changing your string tension, or shifting the racket balance.

Chapter 5

Understanding the Motion of the Ball

(Ball Trajectories)

After the ball has left the racket, the laws of physics take over and determine where it will go, and there is nothing more you or your opponent can do to guide it or change its path. There are three forces acting on the ball during its flight; gravity, air resistance, and—if the ball is spinning—something called the Magnus force, which causes the ball to curve. The force due to gravity is always pointed straight down toward the earth; without it the ball would never end up in the court. Air resistance slows the ball, and in the range of speeds encountered in tennis, the force it causes is proportional to the square of the ball's speed. This means that a ball moving at 60 miles/hour would encounter four times more air resistance force than a ball moving at 30 miles/hour. Table 5.1 gives several examples of how typical groundstrokes are slowed down in flight owing to air resistance. Wind also creates an air resistance force, which can be analyzed in a similar manner. Because air resistance force is proportional to the square of the speed, a crosswind of 20 miles/hour will exert four times as much force on the ball as a 10-mile/hour crosswind, and a 30-mile/hour crosswind provides a force nine times as strong as the 10-mile/hour wind. This is particularly noticeable when you toss the ball up for a serve if there is a brisk breeze. The Magnus (spin) force is at right angles to the direction that the ball is moving and is proportional to how fast the ball is spinning and also to the square of the ball's speed.

A computer program has been written that takes all of these factors into consideration and can turn each one on or off to show its effect. Figure 5.1 illustrates three trajectories computed in this way, first with no gravity (A), then with no air resistance (B), and finally with both gravity and air resistance (C). The shot that is analyzed is a drive hit waist high, from the baseline, with an initial ball speed (as it leaves the racket) of 56 miles/hour and no spin. When the air resistance is eliminated, the ball lands near

Table 5.1. Speed of a Tennis Ball at Various Points in its Trajectory

Initial speed mph	Speed at net mph	Loss of speed mph	Speed at baseline mph	Loss of speed from net to baseline mph
90	74	16	61	13
67	54	13	45	9
45	34	11	27	7

NOTE: The data are for a ball hit from the baseline so as to land close to the other baseline. Its speed is reduced during its flight owing to air resistance. Because the air resistance increases as the square of the ball speed, the ball slows down more between the racket and the net than between the net and the baseline. The faster a ball is moving, the greater its loss of speed.

the baseline. With air resistance, the same flat shot results in a drive that lands well inside the court. Trajectory diagrams will be used extensively to illustrate various points in the material that follows. By using the computer, factors such as gravity and air resistance can be considered or eliminated, and speed, height, angle, and so on can be varied while keeping everything else constant. Such measurements cannot be obtained on a tennis court because no one is a good enough tennis player to be able to vary parameters one at a time in a reproducible manner. It may not seem very useful to analyze trajectory without gravity, but the ability to compute trajectories with varying degrees of air resistance is valuable, because the air resistance is slightly different from day to day owing to atmospheric conditions and is quite different at higher altitudes (such as in Denver, Colorado). In fact, manufacturers produce a high-altitude ball designed to compensate for the higher bounce that the reduced air pressure causes. They do not make a ball that compensates for the reduced air resistance the ball encounters, however, and this is a mistake. For high altitude use, there should be a slightly larger ball (the Wilson Rally size) with a slightly lower bounce (when measured at sea level). Until someone makes this ball, if you plan to play at high altitudes, you should use either the special low-bounce balls or the Wilson Rally balls.

5.1 The Spin of the Ball (in the Air)

When a tennis ball is given a spin (topspin, backspin, or sidespin), its trajectory through the air is modified, and it will dive, float, or

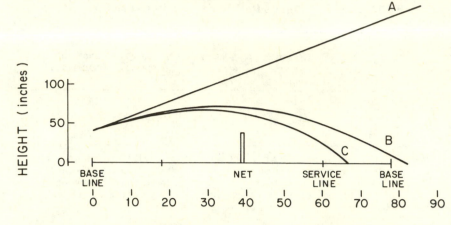

DISTANCE FROM BASELINE (feet)

Figure 5.1. Trajectories of Tennis Balls in Play: Effects of Gravity and Air Resistance. This diagram shows how gravity and air resistance affect the trajectory of a tennis ball. A drive hit from the baseline at an initial speed of 56 miles/hour, from waist height (39 inches), and at an angle of 9 degrees above the horizontal will normally take the trajectory shown by *C*. If there were no air resistance, the trajectory would look like *B* and the ball would land beyond the baseline. If there were no gravity, the ball would never hit the ground (*A*).

curve, unlike a ball with no spin. The Magnus effect causes this modification, and in the same way it allows a pitcher to curve a baseball. The higher the rate of spin of the ball, the more or sharper will be the curvature. You can determine which way the ball will curve (assuming you know which way it is spinning) by this simple rule:

▶ The direction of the spin of the front of the ball is the direction that the ball will curve in the air.

If you hit a ball with topspin, the front of the ball will be rotating downward, so the ball will curve downward (dive or sink). If you chop at the ball to give it backspin or underspin, the front of the ball will be moving up and the ball will tend to stay up or float. If the front of the ball is spinning to the right or left about a vertical axis, the ball will curve or slice to the right or left.

Applying a moderate amount of spin (16 revolutions/second) produces a shot which floats if backspin is applied (curve *A* in Fig-

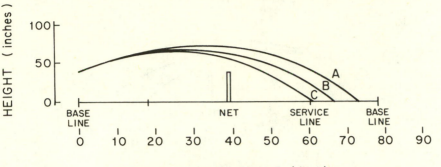

DISTANCE FROM BASELINE (feet)

Figure 5.2. Trajectories of Tennis Balls in Play: Backspin and Topspin (16 revolutions/second) Compared with No Spin. This diagram shows the effect of a spin of 16 revolutions/second on the trajectory of a tennis ball. For a ball hit at an initial ball speed of 56 miles/hour, from waist height (39 inches), at an initial angle of 9 degrees, the trajectories are A, with 16 revolutions/second backspin; B, with no spin; and C with 16 revolutions/second topspin. The topspin shot lands shorter and the backspin shot lands deeper than the shot with no spin.

ure 5.2) or dives if topspin is used (curve C) relative to a flat, spinless shot (curve B). When a larger amount of spin (32 revolutions/second) is used, the ball will either float well out beyond the baseline (curve A, Figure 5.3, for backspin) or it will dive down quite short of the baseline (curve C, Figure 5.3, for topspin) for the same initial ball speed and angle of the ball off of the strings.

Now that we have this computer program, what can we do with it? What questions should be asked? What problems should be solved? There is one problem that comes to mind immediately.

We all make errors when we hit the ball, but some players make barely noticeable ones while others make whoppers. This is one thing that distinguishes good players from hackers and beginners. To a great degree, the magnitude of the error we make is determined by our inherent athletic ability and the amount of practice we are willing to devote to tennis.

We must accept the fact that no player is perfect. Some shots do go long, some do hit the net. This chapter will try to help you improve your game by showing you how to reduce the damage such errors cause. It will show how to play it safe and hit the shot with the largest chance of going in, rather than the shot that has little chance of going in.

By understanding certain physical principles and their con-

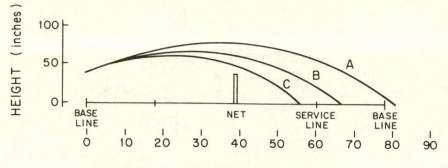

DISTANCE FROM BASELINE (feet)

Figure 5.3. Trajectories of Tennis Balls in Play: Backspin and Topspin (32 revolutions/second) Compared with No Spin. For the same initial conditions as those specified in Figures 5.1 and 5.2, the trajectories with a topspin (*C*) and backspin (*A*) of 32 revolutions/second are compared to that of the shot with no spin (*B*). The topspin shot now bounces even shorter than it did in Figure 5.2, and the backspin shot goes over the baseline.

sequences, a strategy or type of shot can be chosen that will maximize the chances of the return landing in the court. The ball has to go in for the point to be won. What is being advocated here is *high-percentage tennis*.

5.2 Margin for Allowable Error

Whenever you hit a shot, there is a certain margin for allowable error that you must be within, if the shot is to clear the net and not land outside the baseline. You are very likely to bungle a shot if your average error is larger than the shot's margin for error. If, on the other hand, almost every shot you attempt has a large margin for error compared to your own average error, you will be known as a steady player, and you will not be playing a game in which you are beating yourself. Sometimes, of course, you will not want to play the low-risk, high-percentage shot. If the anticipated benefit justifies the risk, you will hit the ball as hard as you can. True, the percentages are against it going in, but if you are successful, the point is yours. What are the odds of getting one of these shots in, compared to a safe shot, and what are the safest shots for you to make?

Angular Acceptance and What It Means to You

Once the ball has left the racket, its trajectory is completely determined by the laws of physics and a computer program can track the ball's flight. Figure 5.4 shows the computer-generated trajectories of two balls that leave the racket at the baseline at the same height, with identical initial speed but different initial angles. The lower trajectory just clears the net. If the initial vertical angle of the ball as it left the racket were any less than the one used in this figure, the ball would hit the net. The upper trajectory bounces just inside the court. If the initial vertical angle of the ball were any greater, the ball would land beyond the baseline. Any initial angle between the minimum angle and maximum angle displayed in Figure 5.4 will result in a good shot. The difference in initial vertical angle between the minimum angle (just clears the net) and the maximum angle (just lands in the court) is called the vertical acceptance angle or the vertical angular acceptance for the particular tennis shot being analyzed. This concept of vertical angular acceptance is very important because it is your margin for allowable error; it tells you what the chances of your getting the ball in are, for any particular shot. If the vertical angular acceptance is large, you have a large margin of allowable error, and you have a much better chance of getting your shot in. If the vertical angular acceptance is very small, say one degree, then unless you hit the ball so it leaves your racket at exactly the correct angle or within a half degree of it, the shot will not go in and you will lose the point.

The same sort of analysis can be used to calculate horizontal angular acceptance from the initial horizontal angle of the ball as it leaves the racket. This will be your allowable margin for horizontal error and will give you an idea of how often you can expect a particular shot to land in the alley instead of the court.

The concept of margin of error (or safe versus risky shot) is an old idea, but using the laws of physics and a computer to put these concepts on a quantitative basis is new. With a computer it is possible to determine how much your chances of getting the ball in change as you move to another spot on the court, hit the ball at a different height, or hit the ball not quite so hard. This book will give you detailed quantitative data on how the various things that you can change will increase or decrease the chances of your shots being good. *If a higher percentage of your shots go in, you will win a higher percentage of the points.*

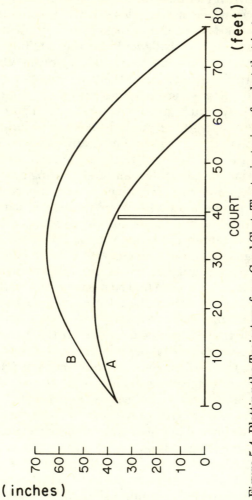

(inches)

COURT

(feet)

Figure 5.4. Plotting the Trajectory for a Good Shot. The trajectories of a shot that just clears the net (A) and a shot that just hits the baseline (B) at a fixed ball speed (67 miles/hour) are displayed here. The difference between the initial vertical angles of the two trajectories constitutes the vertical angular acceptance.

5.3 The Groundstroke

The heart of the game of tennis is the groundstroke, whether forehand or backhand. That is the shot that you must master if you
want to be a successful tennis player. You cannot win without
groundstrokes. Even if you have the best serve and can punch volleys with microprecision, you have to return your opponent's
serve with a groundstroke on every point of every game that you
do not serve. Off the ground, you can hit the ball with either your
forehand or backhand. In doing so, you can hit the ball hard or
soft; you can give it topspin or backspin, or hit it flat; you can aim
it down-the-line or crosscourt. How does each of these options
affect the chances of the ball clearing the net and not going over
the baseline?

Ball Speed

Figure 5.5 shows, for different ball speeds, the minimum vertical
angle that a ball must have as it leaves a player's racket in order
to just clear the net (curve A), and the maximum vertical angle it
can have if it is to land in the court, not over the baseline (curve
B). Several parameters are kept constant: there is no wind, and
the player is hitting the ball each time at waist height, from the
baseline, and with no spin. The only things allowed to change are
the ball's initial speed as it leaves the racket and its initial vertical
angle (relative to the horizontal plane) as it leaves the racket. For
each possible speed that the ball can have leaving the racket, the
angle of the ball that just clears the net and the angle of the ball
that falls on the baseline have been calculated and plotted in this
figure. There are several things of great interest that can be
learned from these two curves.

In Figure 5.5, note that below about 30 miles/hour, curve
A—the just-clears-the-net curve—disappears. This means that
no flat shot hit from the baseline with an initial speed of less than
30 miles/hour can clear the net, regardless of the angle at which
it leaves the racket. Curve B, the just-hits-the-baseline curve, disappears below about 40 miles/hour. This means that no flat shot
with an initial speed of less than 40 miles/hour can possibly go
long regardless of the angle at which it leaves your racket. If a
shot has an initial speed of below 40 miles/hour and clears the
net, it will go in. The difference between the upper curve (B) and

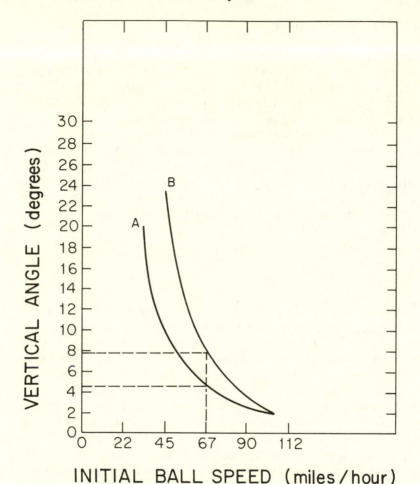

Figure 5.5. Maximum and Minimum Vertical Angles at Different Initial Ball Speeds (No-Spin Shot). The minimum vertical angle (*A*) at which a ball can leave a racket and clear the net and the maximum vertical angle (*B*) at which a ball can leave a racket and land in the court are shown for a range of initial ball speeds (i.e., the speed at which the ball leaves the player's racket). This is for a ball hit waist high from the baseline with no spin.

the lower curve (*A*) is the margin for allowable angular error for a given shot under the constant conditions listed above. Thus, for example, if a player hits the ball at 67 miles/hour (see the dotted line in Figure 5.5), it must have a vertical angle of at least 4.4 degrees to clear the net. If its initial vertical angle is greater than

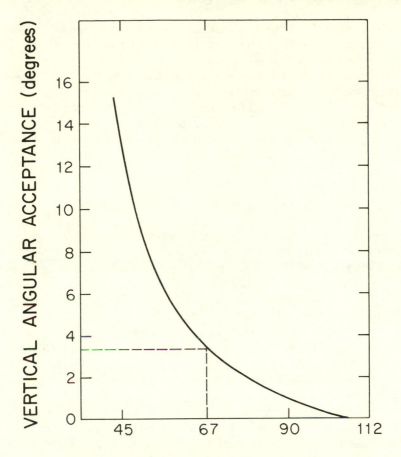

INITIAL BALL SPEED (miles/hour)

Figure 5.6. Vertical Angular Acceptance at Different Initial Ball Speeds (No-Spin Shot). This curve is the difference between curves *A* and *B* in the previous figure.

7.8 degrees, the shot will bounce beyond the baseline. The difference between these two angles (7.8 − 4.4 = 3.4 degrees) is the vertical angular acceptance for a shot hit at 67 miles/hour. This difference has been calculated for all ball speeds and it is plotted in Figure 5.6.

As the ball speed is increased, curves *A* and *B* in Figure 5.5 converge and the difference in angle (which is the vertical angular acceptance plotted in Figure 5.6) goes to zero. You can see graphically how your margin for allowable error decreases as the ball speed increases. The conclusion is clear: The harder you hit the ball, the less likely it is to go in.

Since at 100 miles/hour the two curves in Figure 5.5 meet, the vertical angular acceptance at that speed is zero. This means that you cannot hit a flat (no-spin) shot at waist height from your own baseline at a speed of greater than 100 miles/hour and have it go in, even if your name is Jimmy Connors, Ivan Lendl, or John McEnroe.

Topspin versus Backspin

If you give the ball topspin as it leaves your racket, you greatly increase the chances of that shot landing safely in the court. If you give the ball backspin, you greatly reduce your margin for allowable error. The comparative values of topspin and backspin are illustrated by Figures 5.7 and 5.8, where the trajectories of balls with spins and no spin are shown.

In Figure 5.7 the three balls are hit waist high, at the same initial speed (67 miles/hour), by a player at the baseline. The initial vertical angles of the balls as they leave the racket have been adjusted so that the balls (with topspin at 16 revolutions/second, backspin at 16 revolutions/second, and flat) clear the net by the same height. Note that the topspin shot then bounces quite short and thus gives the largest margin for allowable error relative to the baseline. The backspin shot floats longest and bounces very

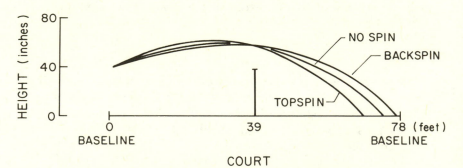

Figure 5.7. Ball Trajectories with Identical Net Clearance. Three shots hit at 67 miles/hour with 16 revolutions/second topspin, 16 revolutions/second backspin, and no spin are shown. The initial angles were chosen so that all three shots would clear the net by exactly the same height.

close to the baseline, thus giving the smallest margin for allowable error. The flat, spinless shot falls somewhere in between.

In Figure 5.8 the three balls are also hit from the baseline
at waist height and with an initial ball speed of 67 miles/hour.
Again the vertical angle that the ball makes as it leaves the
racket is varied, this time so that the ball bounces at the same
location for all three shots (topspin at 16 revolutions/second, backspin at 16 revolutions/second, and no spin). The topspin shot
clears the net by the most and thus gives the player the greatest
margin for allowable error relative to the net. The backspin shot
clears the net by the least and gives the smallest margin for error.

Now that you have seen qualitatively what topspin and backspin can do to a ball's trajectory, let us put even more spin on the
ball (32 revolutions/second) and examine the maximum and
minimum vertical angles for shots that land in the court. Figures
5.9 and 5.10 are the same as Figure 5.5, only topspin and backspin have been applied to the ball. The difference between curve
B (the just-hits-the-baseline curve) and curve A (the just-clears-
the-net curve) is again the vertical angular acceptance or margin
for allowable error for that shot.

In Figure 5.9 (topspin), the difference in maximum and minimum vertical angle is still appreciable at a ball speed of 100
miles/hour (the computer did not carry the calculation beyond
that speed); so if you were strong enough (and skillful enough),
the laws of physics would permit you to make such a shot successfully. In Figure 5.10 (backspin), note how quickly and abruptly
curve A and curve B converge as the initial ball speed increases.

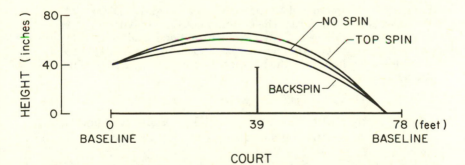

Figure 5.8. Ball Trajectories with Identical Landing Spot. Three shots
with 16 revolutions/second topspin, 16 revolutions/second backspin, and
no spin are shown. The initial angles were chosen so that all three shots
would land at exactly the same location.

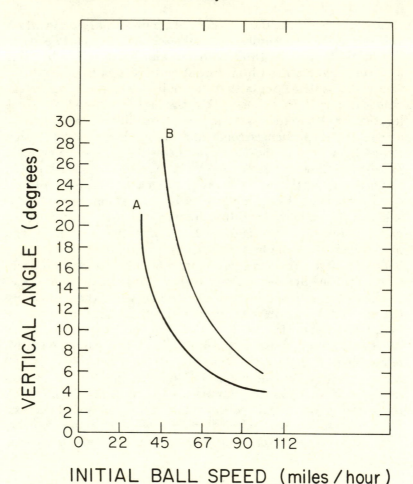

Figure 5.9. Maximum and Minimum Vertical Angles at Different Initial Ball Speeds (Topspin Shot). The minimum (*A*) and maximum (*B*) vertical angles at which a ball can leave a racket and land in the court are shown for a range of initial ball speeds. Curves *A* and *B* are plotted for a ball hit waist high from the baseline with a topspin of 32 revolutions/ second.

Obviously, by putting backspin on the ball a player is decreasing the chance of getting the ball to clear the net and land inside the baseline.

All the data from Figures 5.5, 5.9, and 5.10 have been combined in Figure 5.11. The vertical angular acceptance at each ball speed for each spin condition is displayed. These lines represent the differences between curve *B* (the maximum angle for a good shot) and curve *A* (the minimum angle for a good shot) in the pre-

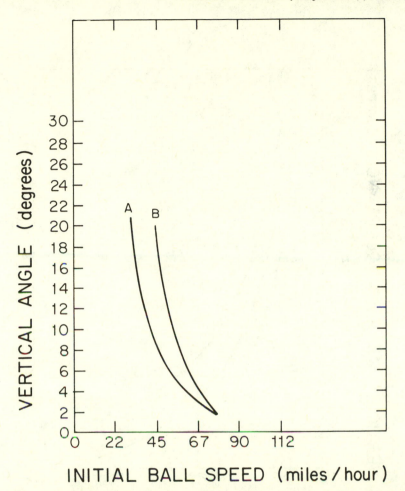

Figure 5.10. Maximum and Minimum Vertical Angles at Different Initial Ball Speeds (Backspin Shot). The minimum (*A*) and maximum (*B*) vertical angles at which a ball can leave a racket and land in the court are shown for a range of initial ball speeds. Curves *A* and *B* are plotted for a ball hit waist high from the baseline with a backspin of 32 revolutions/second.

vious figures. Several points that we have discussed become even clearer in this figure.

▶ The vertical angular acceptance (or margin for allowable error) falls quickly as the ball speed increases. If you want to play safe and just get the ball in, do not hit it hard.

▶ Topspin increases your margin for allowable error, and

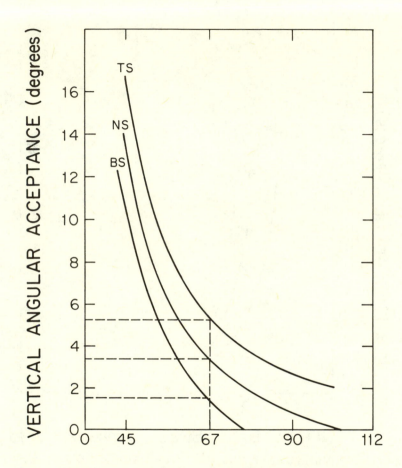

Figure 5.11. Vertical Angular Acceptance at Different Initial Ball Speeds (Topspin, Backspin, and No-Spin Shots). The difference in initial vertical angle between a ball that just clears the net and one that just touches the baseline is plotted here for three kinds of shots: backspin at 32 revolutions/second (curve *BS*), no spin (curve *NS*), topspin at 32 revolutions/second (curve *TS*). These curves represent the respective vertical margins for error for each kind of shot. Thus, by reading upward from, for example, 67 miles/hour (along the dotted line) you can determine the margin for allowable error for each type of shot (1.4, 3.4, and 5.4 degrees respectively) at that initial ball speed.

the relative advantage of topspin increases the harder
you hit the ball.

▶ Backspin reduces your margin for allowable error, and
the relative disadvantage of backspin increases the
harder you hit the ball.

The maximum speed at which you can hit a backspin shot
from the baseline and have it go in, is 78 miles/hour. The maximum speed for a topspin shot to still go in, is off the scale of the
diagram. This means that it is very difficult to hit a good backspin
shot hard. The laws of physics are against you, while they help
you on topspin shots.

For shots hit at 67 miles/hour, the margins for allowable
error vertically for a topspin, a no-spin, and a backspin shot are,
respectively, 5.4, 3.4, and 1.4 degrees (see the dotted lines in Figure 5.11). This means that you have almost four times more margin for error on a 67-mile/hour topspin shot than on a 67-mile/hour
backspin shot, and the flat, no-spin shot has more than twice the
margin for error of the backspin shot. At low ball speeds, the advantage of topspin over backspin is great, but above 67 miles/hour
it becomes even more important.

You do not have to hit your shots with as much topspin as this
figure assumes to gain its advantage. A little topspin will give
you a small increase in your margin for allowable error, while a
great deal of it will greatly increase the margin. All of these data
assume, of course, that you have the same control—that you
make the same degree of error in the vertical angle of the ball as
it leaves your racket—whether you are hitting with spin or flat.
This is usually not the case. If you spray your topspin shots all
over the court but have tremendous control when you chip out a
shot with backspin, then obviously you must use your backspin
strength. Topspin is difficult to learn and is not a shot for the beginner, who should concentrate on the flat shot initially. If good
players want to improve their game, they should learn how to hit
and control a topspin shot, for using topspin is a way to play high-
percentage tennis.

Crosscourt versus the Down-the-Line Shots

When you are returning a shot from the baseline in the corner of
the court, you must decide which is safer, the crosscourt shot or

HORIZONTAL ANGLE

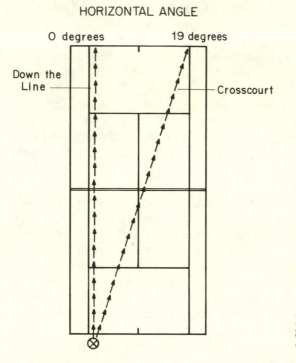

Figure 5.12. Range of
Shots Hit from
the Left Corner

the down-the-line shot (Figure 5.12). In this situation the cross-court shot has a larger vertical acceptance angle than the down-the-line shot and is therefore safer. It is generally accepted that the crosscourt shot is the better choice, but in the many books and articles on tennis that advise you of this, the reason usually given is wrong. They claim that the crosscourt shot has a better chance of going in because the net is lower in the center than at the side-lines. It is true that the net is lower in the middle, but surpris-ingly, this is not the reason why the crosscourt shot has a larger vertical acceptance angle. Computer analysis shows that when you hit a shot of moderate speed from the corner of the court, the minimum vertical angle to just clear the net is almost the same for these two shots. If you are standing in the corner, the center of the net is farther away from you than the edge of the net at the alley, and therefore you must raise the initial trajectory of the ball more to compensate for the extra distance. These two effects, the lower net and the extra distance, almost exactly cancel each other. If the minimum angle to just clear the net crosscourt and down the line is the same, then why is the crosscourt shot more likely to be good? When you hit a ball crosscourt, the baseline is considerably farther away on the diagonal than it is down the line. You therefore have a much bigger court to get your ball to bounce in, and that increases your chances of the shot being good.

All of this is illustrated in Figure 5.13 for a ball hit without spin from the corner with an initial speed of 45 miles/hour. The figure shows the vertical angles for a ball just clearing the net and for a ball hitting just inside the court, for all possible horizon-

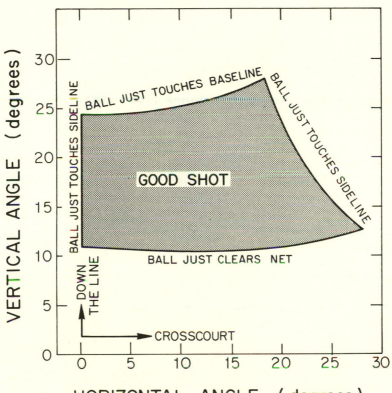

Figure 5.13. Angles of a Good Shot Hit from the Left Corner. This figure shows the angles within which the ball must be hit from the left corner for it to land in the court. The shot is hit waist high at 45 miles/hour. By referring to the previous figure, you can see that if the ball is hit at a horizontal angle of less than zero degrees, it will land in the alley or beyond; or, if it is hit at too large a horizontal angle, it will land in the opposite alley or beyond. The figure also shows the minimum vertical angle for a ball to just clear the net and the maximum vertical angle allowed if the ball is to land in the court, both values being a function of the horizontal angle at which the ball leaves the racket.

tal angles with which you can hit a ball and have it go in. In this figure, a crosscourt shot would be hit at a horizontal angle of between 15 and 20 degrees, while a ball hit down the line would be at zero degrees (see Figure 5.12). You can see that the value of the lower line (the ball-just-clears-net line) hardly changes as you change from down the line to full crosscourt. The value of the upper line (the ball-just-touches-baseline line) increases as the ball angle increases crosscourt. The vertical acceptance (which is your chance of getting the shot in) is the difference in angle between the upper curve and the lower curve in Figure 5.13.

This difference is plotted against horizontal angle in Figure 5.14 and you can see that the crosscourt shot at 19 degrees (A) has more vertical acceptance than the zero degree, down-the-line shot (B). This extra margin for allowable error amounts to about 22 percent for the crosscourt shot over the down-the-line shot when the initial ball speed is 45 miles/hour. This extra margin depends on how hard you hit the ball, however. Figure 5.15 shows you the vertical angular acceptance when the initial ball speed is 67 miles/hour. Note there is quite a difference in the magnitude of the angular acceptance between the two figures. The vertical margin for error is now much smaller than it was at the lower ball speed, but the extra vertical margin of error for a shot hit crosscourt rather than down the line is now 45 percent. At a low ball speed the crosscourt shot was only slightly safer, but as the ball speed increases, the crosscourt shot becomes much safer than the down-the-line shot.

▶ If you are in the corner of the court and you want to hit the ball hard, go crosscourt, not down the line.

This is your prescription for a safe shot when both you and your opponent are at the baseline. The crosscourt shot also puts you in a better position for your opponent's return.

Down the Middle versus the Corners

When you are on the baseline at the center of the court, you can hit the ball down the middle or aim for a corner of your opponent's court. Hitting down the middle, the net is lower and closer, but so is the opposite baseline. When you hit into the corners, the net is somewhat higher but the baseline is farther away. The computer claims that all these factors cancel each other almost entirely, and

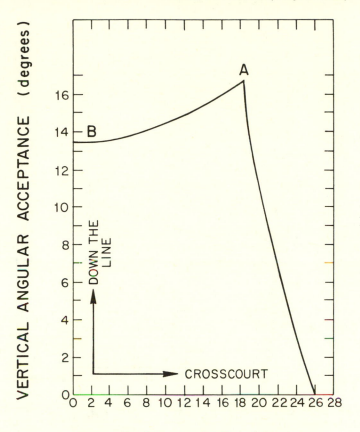

HORIZONAL ANGLE (degrees)

Figure 5.14. Vertical Angular Acceptance versus Horizontal Angle (Hit at 45 Miles/Hour from the Corner). The difference in initial vertical angle between a ball just clearing the net and one just landing in the court is plotted against the actual horizontal angle at which the ball leaves the racket. This figure was derived from the previous figure by subtracting the lower curve (Ball Just Clears Net) from the two upper curves (Ball Just Touches Baseline or Ball Just Touches Sideline) at every horizontal angle. For a horizontal angle between zero and 19 degrees, the limits of vertical angular acceptance are determined by the net and the baseline. For horizontal angles greater than 19 degrees, the limits of vertical angular acceptance are determined by the net and the sideline.

the result is that you have, for all practical purposes, just as much vertical angular acceptance or margin for error down the middle as into the corner. This is illustrated in Figure 5.16; you can see that the vertical margin for error is almost constant.

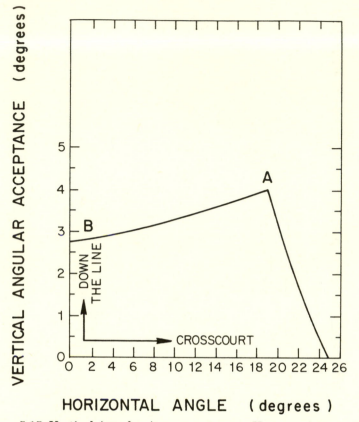

Figure 5.15. Vertical Angular Acceptance versus Horizontal Angle (Hit at 67 Miles/Hour from the Corner). This is the same as Figure 5.14, only the initial ball speed has been increased to 67 miles/hour.

The horizontal margin for error is another matter in this case. Aiming down the center clearly gives you the maximum allowance for error from side to side, while if you aim for the corner, you have very little tolerance for mistakes to the right or to the left.

5.4 Summary

The harder you hit the ball, the smaller is your angular acceptance for that shot going in.

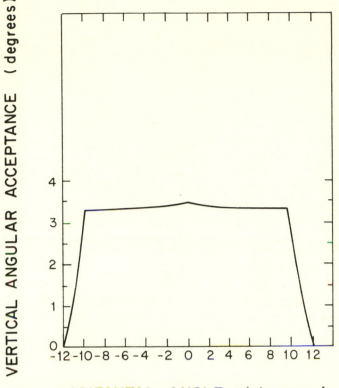

Figure 5.16. Vertical Angular Acceptance versus Horizontal Angle (Shot Hit at 67 Miles/Hour from the Center Line). This is similar to the previous figure, only the ball is hit from the center line instead of the corner. For horizontal angles between plus and minus 10 degrees, the limits of the vertical angular acceptance are determined by the net and the baseline. For angles beyond plus and minus 10 degrees, the limits of the vertical angular acceptance are determined by the net and the sidelines.

Topspin increases the angular acceptance of a shot, which
   means you can hit the ball harder and it will still go in.
Backspin reduces the angular acceptance of a shot.
Backspin tends to make a ball float, while topspin tends to
   make shots dive.
From the corner, a crosscourt shot is safer than a down-the-
   line shot.

Chapter 6

Getting the Ball In (With Position)

The probability of a shot that you hit going in and being good is related to where you are when you hit the ball. If you knew that by hitting the ball from one location the chances of the ball going in were higher than if you hit it from another spot on the court, you would avoid the low probability position and always try to hit your shots from the place that increases the chances of the ball going in (if your opponent is willing to cooperate). You also would aim your shots so that your opponent would be forced to play from a low probability location.

A computer makes just such information available. You will be shown which locations lead to shots having a large probability of going in (a large angular acceptance). These shots may be thought of as being hit into a large court or through a large window. Shots with a small angular acceptance (a low probability of going in) are those hit into a small court or through a small window. When you try to hit into a large court, you have a large margin for error, your shots will go out less often, and you will make fewer errors. When you try to hit into a small court, you will make more errors and lose more points.

What follows is a prescription for defensive tennis. It should be applicable to a large proportion of all of the tennis played in clubs, parks, and resorts. Most points are won when one of the players makes an error, and the player making the fewest errors usually wins the match. For most players, this is winning tennis.

▶ It is very difficult to win points if your shots do not go in.

6.1 Groundstrokes

The further back in the court that you are when you hit a groundstroke, the larger is your probability of getting your shot in, if everything other than distance from the net is held constant. This seems counterintuitive, since you probably feel that, as you move in from the baseline, you can hit the ball harder and it is more

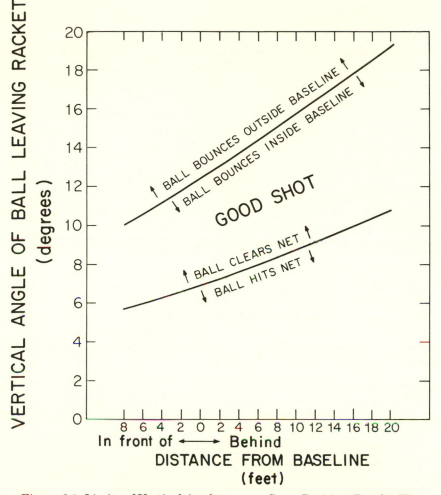

Figure 6.1. Limits of Vertical Angles versus Court Position (Depth). The lower line in this figure is the minimum angle at which a flat (spinless) shot can leave the racket at 56 miles/hour and just clear the net. The upper curve is the maximum angle at which the ball can leave the racket and not go over the baseline if the ball is hit waist high at 56 miles/hour.

likely to go in. Figure 6.1 illustrates that this is just not true. The lower line in the figure is the minimum angle at which the ball must leave your racket if it is to just clear the net for various positions in front of and behind the baseline. (All this is for a ball hit at 56 miles/hour, waist high, with no spin.) As you would expect, the minimum angle to clear the net increases as you move back behind the baseline, since the further back you are, the more you must lift the shot to clear the net. The upper line in Figure 6.1 is the maximum angle at which a ball can leave your racket and not go over the baseline, as a function of where you are on the court when you hit the ball. As you might expect, this angle also increases as you move behind the baseline. What you might not expect is that this maximum angle increases faster than the minimum angle, so that the difference between them (Figure 6.2), which is the vertical angular acceptance, slowly increases as you move back.

Figure 6.2 shows the margin for allowable error or the vertical angular acceptance as you move back behind the baseline for this ball speed (56 miles/hour) and for 67 miles/hour as well. Two things are clear from this figure. First, as has been pointed out before, the allowable error goes down as the ball speed goes up.

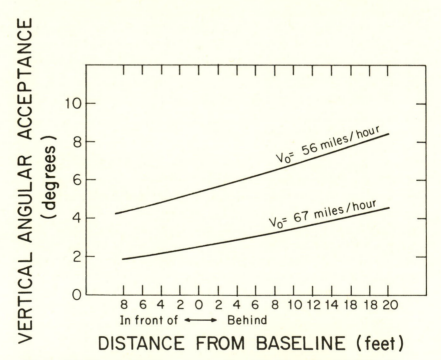

Figure 6.2. Margin for Error versus Court Position (Depth). This diagram shows the difference between the upper and lower curve in the previous figure for two different ball speeds. This difference is the margin for error (vertical angular acceptance) for the shot.

Second, the margin for allowable error goes up as you move behind the baseline. This means you must be careful when you pounce on that somewhat short ball to hit an approach shot, for you are hitting the ball into a small court, compared to the identical shot hit from behind the baseline, and there is therefore more of a chance that you will hit it out. Note this comparison is for the *same* shot. Quite often when you move in to hit the ball, you will hit it at a different height or with a different speed; this comparison is more complicated and will be analyzed shortly. Just remember, the farther back you move, the less critical is the angle at which the ball comes off of the racket, and the safer your shot will be. In addition, as you move back you will have more time to prepare for the shot, and the approaching ball will slow down more because of air resistance. Your errors should thereby be reduced. The one obvious disadvantage to hitting the ball from well behind the baseline is that your opponent will have much more time to prepare to return your shot. You have to make the decision as to whether you want to play a safe shot or move up closer, take the ball earlier, and give your opponent less to work with. This analysis is for a ball hit waist high from near the baseline. If you hit a ball at shoulder height from inside the service box, then you might expect a different result.

Height of Ball When You Hit It (Groundstrokes)

The higher the ball is when you hit it, the larger is the likelihood of that shot going in the court. Figure 6.3 shows the dependence of the vertical angular acceptance on this height for three initial ball speeds. This angular acceptance is the margin for error that you must stay within if your shot is to be good. Many players cannot hit a shot at shoulder height with any power or consistency compared to a shot at waist or knee height. Assuming the control you have over where the ball lands is the same for high and low shots, ball trajectories clearly favor the shot where the ball is hit when it is as high as possible. As an example, a shot hit at 67 miles/hour from a height of 59 inches has twice the chance of going in as exactly the same shot hit at a height of 20 inches. At 78 miles/hour, the difference in angular acceptance between hitting a ball when it is 59 inches high and hitting it when it is 20 inches high is a factor of four. This means that you are four times less likely to make an error that will lose the point if you

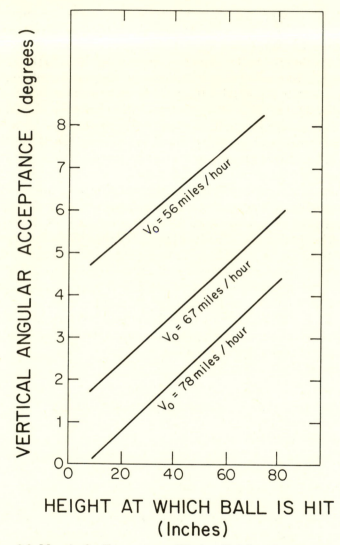

Figure 6.3. Margin for Error versus Hitting Height. This diagram shows how the margin for error increases as the height at which the ball is hit increases (for three different initial ball speeds).

hit the ball when it is high, assuming that your control is the same for both shots.

▶ If you want to increase the chances of your shot going in, hit it when the ball is high—at the peak of the trajectory, if possible.

The Real World

In the above calculations, either the distance from the net or the height was kept constant. Tennis is not played this way, however. On every shot there is a continuously changing correlation between the ball's height and its position on the court. If you move back, you gain a little because of your position, but you usually lose more because you have allowed the ball to drop. What is clear is that if you want the safest shot, you should never hit a ball while it is still on the rise. Hitting a rising ball is for advanced players.

If you can hit a high ball just as well as a ball down around your knees, to increase your chances of a shot going in, you should move so as to hit the ball when it has reached or just passed the peak of its trajectory. This concept is illustrated in Figures 6.4

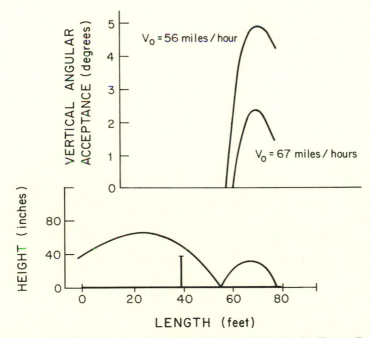

Figure 6.4. Ball Trajectory, Court Position, and Margin for Error (Return of a Groundstroke). The lower part of this figure shows the trajectory of a typical groundstroke. The upper part of the figure shows the margin for error (vertical angular acceptance) for returning this shot as a function of where you are when you hit the ball. The two upper curves are for returns of 56 and 67 miles/hour. You obtain the maximum margin for error when you are slightly farther back than the peak of the trajectory of the incident groundstroke.

and 6.5. The lower part of each figure gives the actual trajectory
of your opponent's shot. In Figure 6.4 this curve shows a groundstroke that bounces quite short in your court. Directly above it is
plotted the vertical angular acceptance for your return as a function of where you are when you hit the ball. This takes into account the effects of both the height of the ball and your position.
The two curves in the upper section of this figure are for return-
ball speeds of 56 and 67 miles/hour. Comparing the upper two
curves with the lower curve shows that the angular acceptance of
your return shot is greatest just slightly behind where your opponent's shot reaches the highest point in its trajectory. In other
words, when you return such a shot, you must move in and catch
the ball near its peak and not wait for the ball to come to you. Any
other strategy reduces the chances of your return being good.

Figure 6.5 is for a serve by your opponent; the ball is still
rising after the bounce when it crosses the baseline (lower curve).

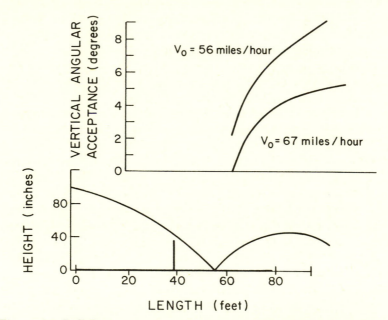

Figure 6.5. Ball Trajectory, Court Position, and Margin for Error (Return of a Serve). The lower part of this figure shows the trajectory of a
typical serve. The upper part of the figure shows the margin for error
for returning the serve as a function of where you are when you hit the
ball. The two curves are for returns of 56 and 67 miles/hour. The deeper
behind the baseline you stand, the better are your chances of returning
this serve, if you were only to consider your margin for error (vertical
angular acceptance).

The two curves at the top of the figure represent the vertical angular acceptance or margin for error for your return of your opponent's serve at 56 and 67 miles/hour. Both curves continue to increase as you move back well behind the baseline, and therefore the chances of your return being good continue to increase. In addition, as you move back, because air resistance slows the ball somewhat and the ball must travel further to you, you have extra time to prepare your shot. Your chances of returning the serve successfully are thus even greater than the analysis indicates.

6.2 The Volley

The distance you are from the net and the height of the ball when you hit it are the most important factors in determining whether you have a good chance or only a marginal chance of getting your volley to go in. Once these are determined, the only thing that you can do to increase the likelihood of your shot going in, besides hitting the ball in the center of your racket, is to vary the speed of your return. When you are close to the net and the ball is well above it, your margin for allowable error is so large that you should not have to worry about the success of the shot.

This is shown in Figure 6.6, which gives the angular acceptance for vollies hit at 45 miles/hour at various heights above the ground and for various locations in the court relative to the net. If you hit the ball when it is 3.9 feet above the ground (about 10 inches above net level), your vertical angular acceptance is huge if you are within a few feet of the net. As you move back from the net, your acceptance margin of error decreases to about 10 degrees, which is still fairly large. As you can see, the lower the ball is when you volley, the smaller is the window through which you must hit the ball to be successful. If you are going to hit the ball below the level of the net, then your chances of hitting a good shot increase as you move back from the net. A ball that you volley when it is 2 feet above the ground (one foot below net level) will not go in if you are within 9 feet of the net and attempt to hit it so it leaves your racket at 45 miles/hour.

Figure 6.7 shows how your chances of a successful volley depend upon how hard you hit the ball if you meet the ball 3½ inches above the level of the net. The two curves represent the vertical angular acceptance when you are 40 and 80 inches from the net. You can see how fast your window closes as you try to hit your volley harder. Under these conditions, a volley leaving your racket

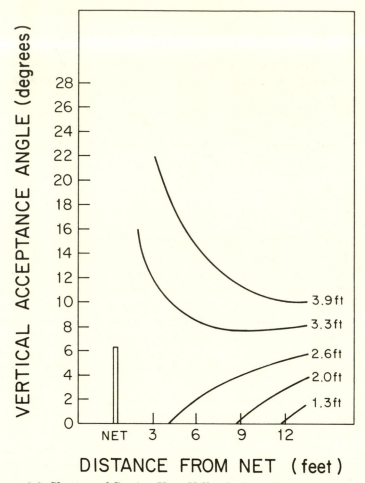

Figure 6.6. Chances of Getting Your Volley In (As a Function of Court Position and Ball Height). This diagram shows the margin of error for a volley as a function of your distance from the net for five different hitting heights. The ball leaves the racket at 45 miles/hour.

at 45 miles/hour is over 4 times as likely to be good as a volley hit at 78 miles/hour.

The Real World (Volley)

The preceding section showed the angular acceptance of volleys as a function of the distance from the net for several different but constant heights at which the ball is hit. This is clearly an idealiza-

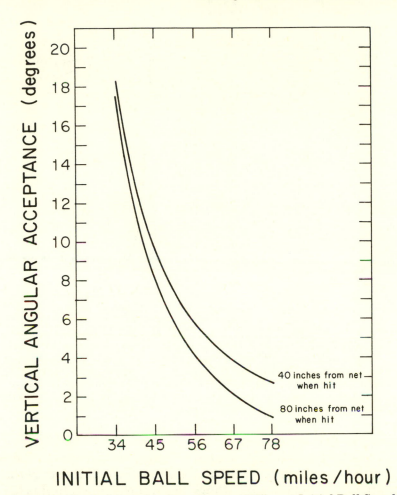

Figure 6.7. Vertical Angular Acceptance at Different Initial Ball Speeds (The Volley). The vertical angular acceptance of a volley that is hit 3½ inches above the level of the net is plotted against the speed of the ball as it leaves the racket for two different distances from the net (40 inches and 80 inches).

tion, since as you move closer to or away from the net for a given shot, the height of the ball changes.

Figure 6.8 brings this all into perspective. The upper part of this figure shows the trajectory of two shots hit by your opponent from the baseline, a flat drive and a topspin shot, both of which have the same clearance of the net. The lower half of the figure shows the angular acceptance of the return of these two shots as a function of the distance from the net when the ball is hit. The

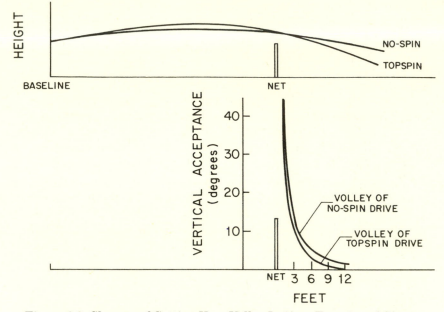

Figure 6.8. Chances of Getting Your Volley In (As a Function of Court Position and Ball Height). The upper two curves are the trajectories of typical groundstrokes hit from the baseline at a speed of 67 miles/hour with no spin and with topspin. The bottom two curves show the vertical angular acceptance (margin for error) for a volley attempting to return those groundstrokes at a ball speed of 56 miles/hour.

baseline drive was hit at 67 miles/hour, while the return volley was hit at an initial ball speed of 56 miles/hour. Several things are quite clear from this diagram. The angular acceptance of the return volley is enormous within a few feet of the net. In this figure it falls off much faster as you move away from the net than it does in Figure 6.5 because the ball is allowed to follow its true trajectory while in Figure 6.5 it is kept at constant height. The topspin shot is clearly more difficult to return because it dips below the level of the net much faster than the flat shot. In fact, if you are more than 10 to 12 feet from the net, it is impossible to volley the topspin drive with any reasonable pace; you still have a few degrees angular acceptance for a 56-mile/hour return of the flat drive, however.

The standard advice that one gets is, "close into the net on a volley"; the closer you are, the better off you are. Do the results of computer analysis agree? The data presented in the figures graphically shows your margins for error. For a volley that you hit well above net height, there is no doubt that the closer you are to the net, the larger your angular margin for error is, both vertically and horizontally. That means that the ball is much less likely to go long or into the net if you crowd the net, and you can get much sharper horizontal angles from in close. Once the ball drops below the level of the net, however, you have a problem if you are in close. If the ball drops low enough, it may not be possible to return the shot with any pace, and you may have to just drop it in or lose the point. This large degradation of your margin for error as you approach the net for a low shot is not real, however. You are not likely to get a very low shot when you crowd the net, because the ball has to cross over the net, and it does not suddenly drop unless it is hit with a great deal of topspin. You are much more likely to be faced with a very low volley, or even a half volley, when you are 12 feet from the net than when you are 6 feet from it.

6.3 The Half Volley

It should now be obvious why the half volley is such a difficult shot. You must hit the ball when it is very low, and you usually hit it from well inside the baseline. As you have seen, these two conditions reduce the chances you have to get the ball in the court because your vertical acceptance angle is reduced to almost nothing. The only way to increase your chances of making a successful half volley is to hit the ball with very little pace. Unfortunately, your opponent may then pounce on it and hit a winner. It is thus a losing position.

There is, of course, another difficulty in the half volley—the problem of timing and the bounce—which makes the shot basically a block, rather than a swing or stroke. The larger-head rackets are an advantage in such a situation.

The conclusion is obvious:

▶ If you play two half volleys in succession, you are foolish.

Looking at this from the other point of view, it is clear that you should try to hit your shots so that they bounce at your opponent's feet. This will mean that the return will be a low-percentage, low-margin-for-error shot that usually will not have any pace on it.

6.4 The Serve

Tennis lore is full of stories about the trajectory of a serve; such claims as, "You have to be 9 feet tall to be able to get a serve in with no spin," are common. To investigate the validity of such assertions, the computer was programmed to calculate the trajectories of serves under several conditions. It turns out that you do have to hit the ball at least 9 feet above the ground if you only consider geometry and omit the effects of gravity and air resistance on the ball. Under the conditions you face in the real world, however, the computer says that you can get a serve into the box, even if you are not 6 feet tall.

The Height and the Serve

There is a strong correlation between the height at which the serve is hit and the chances of that serve being good. The curves in Figure 6.9 correlate the height of the ball when hit with the vertical acceptance angle (margin of allowable error) for serves of 67, 90, and 112 miles/hour. As you can see, the chances of hitting a good serve go down very quickly as the ball height decreases. As an example, if you hit your serve at 90 miles/hour, increasing the ball height from 85 to 105 inches will double your chances of getting the serve in. The harder the serve is hit, the more important the height is. For a 67-mile/hour serve, the increase from 85 to 105 inches in hitting height will only improve the serve's chances of success by 35 percent. At 90 miles/hour the increase is 100 percent, and at 112 miles/hour, the effect of height is even more spectacular. Note that if you hit a flat, 112-mile/hour serve from a height of 80 inches, it cannot go in; at 85 miles/hour it is unlikely to go in, but possible. At this ball speed, the increase in height from 85 to 105 inches will increase your chances of a good serve by a factor of four (400 percent).

There is no doubt that tall players can have a great advantage on the serve. You clearly do not want to choke up on the racket handle when you serve, because that can only lower the point at which you hit the ball.

Ball Speed

Many players love to hit their first serve as hard as they can; they then hit the second serve at a moderate speed, since the cannon-

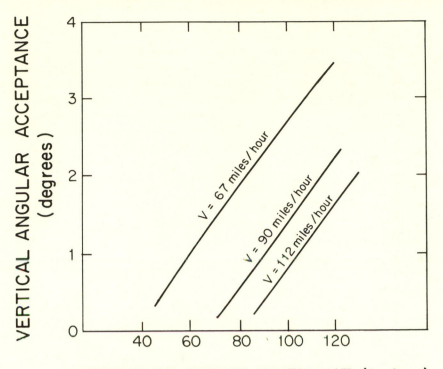

Figure 6.9. Chances of Getting Your Serve In (As a Function of the Height of the Ball when Hit). The vertical angular acceptance is plotted against the height at which the ball is hit, for three different ball speeds of a flat (spinless) serve down the middle.

ball rarely goes in. The subject of serving strategy is covered in a later chapter, so only the ball trajectories and their margins for allowable error are examined here.

Computer analyses show, not surprisingly, that the harder you hit your serve, the more difficult it is to get it in. This is shown in Figure 6.10, where the vertical angular acceptance or margin for allowable error in making a serve is plotted against initial ball speed for two different ball heights. This figure shows that if a player who hits the ball at a height of 85 inches reduces the ball speed to 90 miles/hour from 112 miles/hour, the probability of the ball going in will increase by a factor of more than three. The same decrease in ball speed for a tall player who hits the ball when it is 105 inches above the ground will only increase the chances of the ball going in by 50 percent. Looking at these numbers in a different light, if your 90-mile/hour serve is

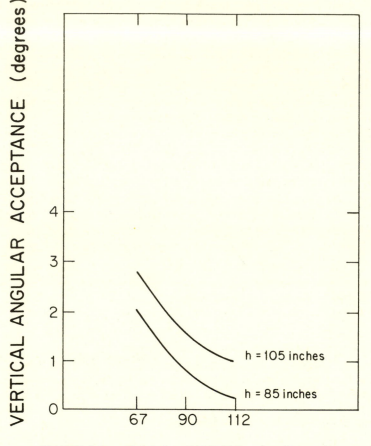

Figure 6.10. Chances of Getting Your Serve In (As a Function of the Ball Speed). The vertical angular acceptance is plotted against the initial ball speed off the racket for two different heights at which the ball is hit.

successful half (50 percent) of the time, then your 112-mile/hour serve will succeed only a third to two-thirds of that amount (16 to 33 percent of the time), depending on whether you are short or tall.

Topspin on the Serve

Topspin will give you a much bigger margin for error on your serve, allowing you to get a much greater percentage of your serves in the court. Figure 6.11 shows the trajectories of two serves, one with topspin and one with no spin, at the same ball speed. The topspin serve crosses the net with a greater clearance and then bounces shorter than the no-spin (or flat) serve. Therefore, there is a larger margin for error at both the places that are crucial in getting your serve in, the net and the service line. Figure 6.12 shows the difference between these shots quantitatively. Here the vertical angular acceptance is plotted as a function of ball speed for four different serves—a topspin serve hit at 85 and 105 inches above the ground and a serve hit without spin at the same heights. The important message in these four curves is simple: a short player who uses topspin can get more serves in than a tall player who uses no spin whatsoever. The percentage increase in vertical acceptance angle is so large that anyone who has difficulty getting in serves with good pace should consider using some topspin.

▶ If you serve the ball hard, a little topspin can easily double your chances of hitting a good serve.

Your Position on the Court for the Serve

When the position of the server is varied the results are very interesting. The top players inch up to the baseline as if every

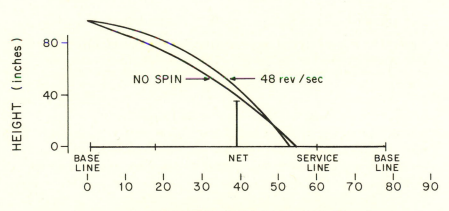

Figure 6.11. Trajectories of Serves. The trajectories of two serves hit from an initial height of 105 inches, with an initial ball speed of 78 miles/hour, with and without topspin are shown.

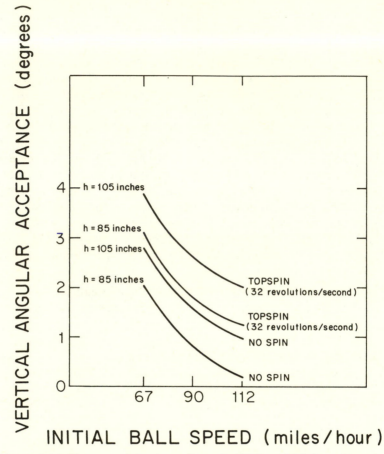

Figure 6.12. Chances of Getting a Serve In With and Without Topspin. The vertical angular acceptance of four different serves is shown as a function of the initial ball speed off the racket. The four conditions are with topspin from heights of 85 and 105 inches and without topspin (flat) from the same two heights.

micron made a difference. As you can see, Figure 6.13 shows that it does make a small difference to them in getting their serve in, but it may not make a difference to you unless you are tall or hit the ball hard. The two heights used in the calculation, 105 and 85 inches, were derived by measuring the heights at which six-foot and five-foot players hit their serves. If you are short or do not hit the ball at a point where it is as high as you can reach, and you do not hit your serves hard (less than 70 miles/hour), the margin for error for your serve may even increase very slightly as you

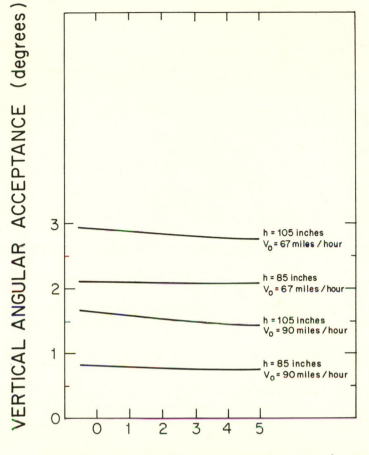

Figure 6.13. Chances of Getting Your Serve In (As a Function of Your Court Position). This figure illustrates how the chances of your being able to get a serve in the service box change as you increase your distance behind the baseline (for two different hitting heights and two different ball speeds).

back away from the baseline. This is because the ball will slow down more in flight owing to air resistance and will fall faster as it crosses the net. Of course, as you move back, your serve becomes weaker, since your opponent has more time to see it, prepare for it, and return it. The point is that unless you have a very strong serve (90 miles/hour or more), there is no need to stand within a quarter of an inch of the line—it gains you little or

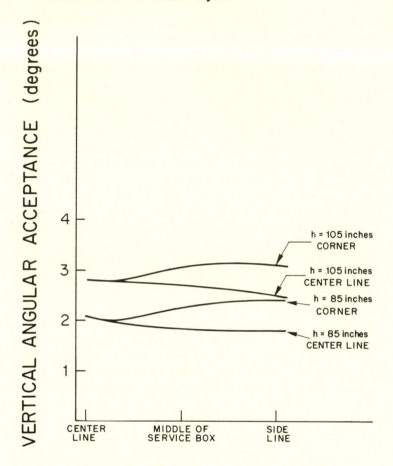

BALL BOUNCE POSITION

Figure 6.14. Chances of Getting Your Serve In (As a Function of Ball Destination Point). This diagram shows the chances of getting a serve in the court as a function of where the ball bounces—near the center line, near the alley, or in the middle of the service box. Serves originating from the center and the alley corner, hit from two different heights, are displayed.

nothing and it may cost you foot faults. Even with a 90-mile/hour serve, moving one inch will change the chances of your serve going in by only a third of a percent (one time in 300). Since you serve only 100 times in an average three-set match, standing one inch closer will allow you to get in only one additional first serve in every three matches. If you hit your first serve at 112 miles/

hour and it goes in with some regularity, you are probably on the pro tour, and the inch closer will give you one extra cannonball serve that is good per 100 attempts. For you, then, this one extra first serve that is good per three-set match, at 112 miles/hour, may be significant.

There is another direction that you can move in choosing a serving position—sideways. In singles, the server normally stands fairly close to the center; in doubles, positions range from the center to the corner. How does this change of position affect the probability of getting a serve in? Surprisingly, Figure 6.14 shows that you have a 10 percent larger probability of getting a serve in from the corner for most of the range of angles that are accessible to you. This means that in singles play you will get in one extra first serve every other game, or three extra serves per set. If you want to serve right down the middle, you are better off standing in the center of the court, but if you want to serve wide and pull your opponent off the court or jam your opponent by aiming at where he or she is standing, the margin for error favors a serve from the corner.

There are several disadvantages of serving from the corner. In singles, your opponent has extra time to hit the return of a serve delivered from the corner, compared to the time that is available for serves from the center line, and you are in a bad position to get that return. In doubles, the second reason does not hold, so you may position yourself depending upon what you are trying to accomplish with the serve. If you want to force your opponent to return the ball from outside of the alley, then standing near the alley yourself, increases both your chances of the serve going in and the angle of the serve. This increase in angle will pull your opponent even farther off the court. Computer studies show that moving to the corner enhanced the chances of getting a serve in even more for a short player than for a tall player.

6.5 Summary

Groundstrokes.

Moving back increases your chances of your shot going in.
Hitting the ball when it is higher increases your chances of the ball going in.
If you can move up to hit a ball at the peak of its trajectory, do so.

Volley.

>Closing in toward the net to volley increases your chances
>of getting your shot in.

Serve. You increase your chances of the serve going in when you

>increase the height at which you hit the ball
>decrease the ball speed off your racket
>increase the topspin on the shot
>position yourself closer to the alley

Chapter 7
Getting the Ball In (Using Your Strokes)

Despite all the theory and computer analysis, you still have to hit the ball yourself. Choosing a good strategy and position, hitting high-percentage shots, and using the proper equipment may help you win more points, but basic strokes still dominate the game, and you may not win the match if you cannot place the ball where you want it to go as well as your opponent can. Even players who are able to aim the ball fairly well may lose their control when their opponent starts to put pressure on them. What causes this? Can a knowledge of the physics of tennis help to improve control?

7.1 Reducing Side-to-Side Errors on Groundstrokes

Old books on tennis generally tell you to "open" your stance (face the net) a little in order to hit a ball crosscourt, and to "close" your stance (stand sideways to the net) somewhat in order to hit down the line. This is the normally accepted way of hitting shots. (In fact, some people try to anticipate where their opponent is going to hit the ball by reading his or her body position before the ball is hit.) Actually the stance does not determine where the ball goes. What gives the ball its direction is the angle of the racket face and the direction of the racket velocity at the instant of contact between the ball and the racket. For the purpose of the analysis in this section, let us put aside all strokes with spin and assume that the racket face is oriented perpendicular to the direction of the racket motion (i.e., to produce a normal, flat drive). The horizontal direction of the ball as it leaves the racket is then determined predominantly by timing. If you swing a little early, or a little late, or if you misjudge the incoming ball's speed, your shot will end up in the alley instead of in the court.

The Problem

When you stand at the forehand corner of the court (as shown in Figure 7.1) and try to return a shot to the center of your oppo-

EARLY   PROPER   LATE
SWING   SWING   SWING

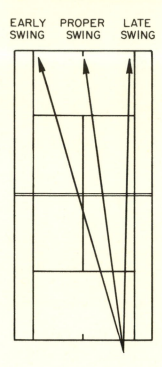

Figure 7.1. Shots Hit from the Corner of the Court. This illustrates shots aimed at the center of the court when you have hit the ball early, late, or correctly.

nent's court with a forehand drive, the shot will go crosscourt if you swing a little early, or down the line if you swing a little late. Thus, most of your shots will still go in. If you had, instead, aimed down the line or crosscourt, some of the shots would have gone wide.

The swing of a tennis racket can be described as the arc of a circle. At the instant you hit the ball, your racket is in a certain position in the arc. Consequently, the face of the racket is pointing in a certain direction, and at that instant the racket is moving tangent to this arc. From Figure 7.2 it is clear that if you hit the ball early or late, the racket orientation and the direction of the racket motion at the instant of impact will be different than they would be if the ball were hit at the proper time. The actual angular error of the racket is given (in degrees) by the formula:

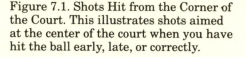

57 × timing error × (ball speed + racket speed) / swing radius.

This means that the worse the timing error, the larger the angular error. This error decreases as the swing radius increases (this will be discussed shortly), but it increases as the racket speed and the speed of the approaching ball increase. This explains

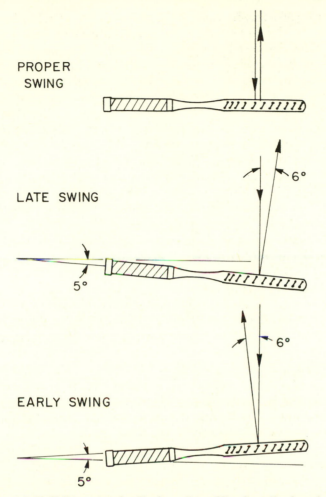

PROPER
SWING

LATE SWING

6°

5°

6°

EARLY SWING

5°

Figure 7.2. Ball Direction for Early or Late Hits for a Racket that is Off
by 5 Degrees

why your control may seem to fail somewhat as your opponent puts
pressure on you (the ball speed coming at you is increased), and
why you should hit the ball softly to a beginner who is learning
the basic strokes. It also explains why you tend to have control
problems when you hit the ball a bit harder (the racket speed is
increased).

If the racket does not move, but merely blocks the ball, and
its angle is in error by 5 degrees, the direction of the ball will be
off by 10 degrees. But if the ball is hit with good pace, this angle of
ball error is reduced from 10 degrees to about 6 or 7 degrees

(when the racket error angle is 5 degrees). For a shot hit from the baseline, the horizontal error in ball location (in feet) when it reaches the other baseline will be the error in the angle of the ball coming off the racket (in degrees) times 1.36. For example, if the ball leaves the racket at an angle of 6½ degrees from its aimed direction, the ball will be 9 feet from where you were aiming it by the time it gets to the other baseline. The singles court is only 27 feet wide, so an error that will put the ball 9 feet to either the right or the left of where you intend it to land is an error of two-thirds of the court's width. With this sort of error in your shots, you must aim for the center of the court and forget about trying to hit the ball down the line or crosscourt. If your error in racket angle is appreciably greater than 5 degrees, you will be spraying your shots all over and out of the court.

Let us take an example to show what sort of timing accuracy you need if you want only a small error in ball placement and how you can use physics to overcome this limitation. Assume the racket head is moving at about 60 feet/second and the ball is approaching at a comparable speed (60 feet/second is close to 40 miles/hour). If you swing the racket with an arc so that the impact point of the ball on the strings is 3 feet from the center of the arc (a very wristy swing), an error in timing of 0.01 second corresponds to an angular error of 11 degrees in racket position, which is a 14-degree error in ball direction. This becomes a 19-foot change in the ball's position by the time it reaches your opponent's baseline. This is clearly not acceptable, since the alley is only 13.5 feet from the center of the court.

To try to reduce this variation in where the ball lands, you might try to reduce your timing error. But there is very little that you can do to improve your timing beyond a certain limit, and you may be close to that limit now. The error used in the example— 0.01 second—is hard to achieve, let alone improve.

Swinging slower surely would reduce this variation, since the error you make increases in direct proportion to the racket-head speed. If you use a racket with more power (i.e., a higher co-efficient of restitution), you will not have to swing it so fast to get good ball speed, which will reduce this sort of error and produce other beneficial effects. That is one of the reasons why racket manufacturers stress the power you can get out of their racket and why this book tells you about the coefficient of restitution.

A more powerful racket—one with a higher COR—will give you higher ball speeds with less effort on your part; it could also give you slightly better control because you do not have to swing the racket so fast.

The Solution

Increasing the radius of your swing will improve your accuracy and control. To do this, you must follow the standard instructions—"keep the ball on the racket as long as possible," "follow through," "keep your wrist firm."

A wristy swing—that is, one in which the wrist is the effective pivot about which the racket head rotates or arcs—has a very short radius. This kind of swing gives you poor control and will result in large errors. If you keep a firm wrist and use your shoulder as the pivot point for your shots, you will double the radius of your swing (the distance from the ball impact point to the pivot) and reduce by half the horizontal angular error caused by the timing error associated with that shot. You can reduce this angular error even further if you try to swing, not in an arc, but in an almost straight line by making the racket head follow the path that you intend the ball to take. This means that your follow-through should be in the direction of the desired ball flight, which can be accomplished by keeping the wrist laid back, moving the whole body along with the shot, and swinging so as to keep the ball on the strings as long as possible. If the orientation angle of the racket head and the direction of the racket's motion do not change over a large portion of the swing, then you have much more control over the direction of the ball's flight. If you do not change the racket angle over the entire swing, then you have almost eliminated the timing error as a source of your placement error. This is illustrated in Figure 7.3; the error in the angle of the ball decreases as the radius of the swing increases, while the timing error (0.01 second) and swing speed remain unchanged. While you cannot move the racket in a straight line over the entire swing and get any power or flow to your shot, you can try to keep the racket moving in a single direction over that region of the swing when the ball is in contact with the racket.

▶ The larger the radius of your swing, the less timing errors will affect you.

Many of the top players do not hit this way because they want to hit the ball hard; they must use some wrist or pronate the forearm to get that extra snap they need to win points and apply pressure to their opponents. They can afford to do this because their timing is very good. The average player who wants more power but cannot afford to give up control should use a racket with a higher COR.

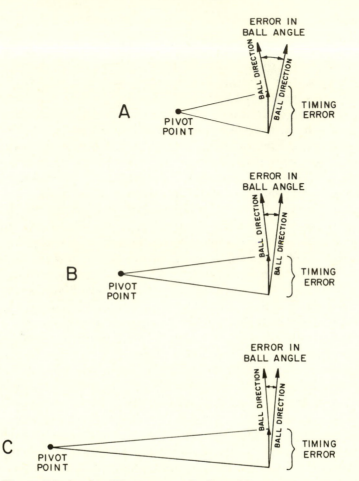

Figure 7.3. Angle of Error versus Radius of Swing. A given error in the timing of your swing corresponds to an error in the location where your racket meets the ball. For a given error in the location, the angle at which the ball leaves the racket is a function of the swing radius. The larger the swing radius, the smaller the angular error. These three diagrams correspond to effective pivot points of the swing in the wrist (A), shoulder (B), and actually behind the body (C).

Increasing the string tension reduces the dwell time of the ball on the strings, which means that the angle through which the racket moves while the ball is in contact with it is reduced. This should, in theory, increase your ability to control the placement of your shots. In practice, however, this difference of dwell times (a few thousandths of a second) is usually much less than the error in your timing, so your accuracy may not be significantly improved.

The Blocked Shot (or Chip)

On the return of a very fast ball—for example a serve, when the ball speed can reach 100 feet/second—a slight timing error can be magnified into a large error in angle. Many players therefore chip, or block, the ball, using the ball's own energy to generate the return speed. In a chip or a block, the player does not swing at the ball in an arc, but punches it, trying to hold the racket-head orientation constant. A small timing error in this type of shot has little effect, compared to what would occur in a normal swing. The block may not be a pretty or graceful shot, but it gets the ball back where you want it (usually just in, on the return of a 100 feet/second serve), and this is what wins points.

Tennis is not Baseball

A tennis swing is not like a baseball swing. If your racket ends up with the face aiming at right angles to the ball direction because of the follow-through, you will have problems placing the ball where you want it to go. Instead of using the baseball swing, with its roundhouse motion, try to keep the racket face open (toward the ball direction) and end the swing high with the follow-through.

Because the ideal baseball swing is level (the bat moves in a horizontal plane), there are many foul balls. Unlike baseball, however, in tennis you lose the point when the ball lands outside the line. You should therefore try to keep the racket moving in the direction that you want the shot to go, both before and after you hit the ball.

7.2 Errors in the Horizontal Direction Because You Do Not Know Which Way to Aim the Ball

The preceding section gave examples of a ball coming into the racket at an angle of 5 degrees and leaving it at just one or two degrees beyond the racket direction, not the five degrees that you might expect. Isn't the angle of reflection (or rebound) equal to the angle of incidence? That rule only holds (and then only approximately) for a racket that is not moving. The incident and rebound angles of a tennis ball off a stationary racket are approximately equal, because the COR of the racket (about 0.6) is very similar to the ratio of speed of a ball parallel to the racket face after being hit to its speed before being hit.

To calculate the size of the rebound angle from a racket in motion, you must first move into the frame of reference where the racket is at rest, compute the angle, and then determine the angle in the frame of reference of the tennis court. The results of this kind of computer analysis are described in this and the next several sections.

The Angles

When a ball hits the face of a racket at an angle that is not perpendicular to the face, the angle that it leaves the racket depends on how hard the ball is hit. The higher the speed of the racket head, the closer the ball will move in the direction that the racket is moving. This is illustrated in Figures 7.4 and 7.5.

This means that when you try to return a crosscourt shot down the line, if you swing the racket so it is moving exactly down the line, the ball will go in the alley if you hit it hard, and outside the alley if you do not hit it with good pace. A crosscourt shot comes toward you at 15 to 20 degrees relative to the sideline. If you aim your racket right down the line, as is shown in Figure 7.5, the ball will go wide—the exact angle is determined by how hard you hit the ball. For example, if the ball comes at you at about 60 feet/second and you merely block the shot, the ball will leave your racket at 15 to 20 degrees and bounce in the next court. If you swing your racket at 20 feet/second, the ball will leave the head at 60 feet/second but at an angle of 10 degrees, and it will land beyond the alley. If you swing at 60 feet/second, the ball will leave the racket with a speed of 118 feet/second at an angle of 5 degrees and will land just outside the alley. To get the ball in the court but down the line, you must have the racket head pointing at a spot at least one alley width inside the sideline if you hit the ball hard, and two alley widths inside the sideline if you do not put a great deal of pace on the shot.

The same strategy can be used when you try to return a down-the-line shot crosscourt. Your racket must point at least one alley width inside the opposite sideline if the ball is to bounce deep and in the court.

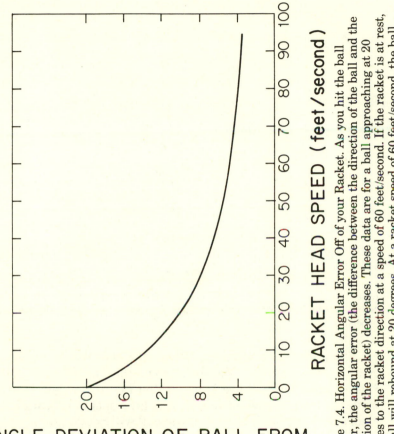

RACKET HEAD SPEED (feet/second)

ANGLE DEVIATION OF BALL FROM
RACKET DIRECTION (degrees)

Figure 7.4. Horizontal Angular Error Off of your Racket. As you hit the ball harder, the angular error (the difference between the direction of the ball and the direction of the racket) decreases. These data are for a ball approaching at 20 degrees to the racket direction at a speed of 60 feet/second. If the racket is at rest, the ball will rebound at 20 degrees. At a racket speed of 60 feet/second, the ball will leave the racket at 5 degrees to the racket's direction of motion.

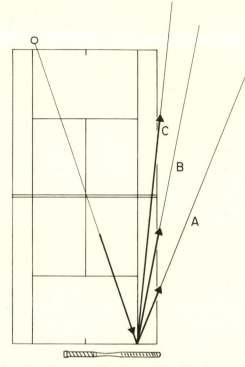

Figure 7.5. Angular Error on the Court. The harder you hit the ball, the smaller the angle between the direction the ball goes and the direction of the racket-head motion. The length of the arrows *A, B,* and *C* indicate the speed of the ball hit by a racket aimed down the line. It is clear that the harder the balls are hit, the closer they go to the racket direction.

When you try to return a crosscourt shot with a crosscourt shot, or a down-the-line shot back down the line, the ball will go where you aim it because it meets the racket head-on (perpendicular to the face). When you want to hit a shot to a location that is not where the ball came from, the ball will not go exactly in the direction of the racket motion; you must therefore swing at an angle that is determined not only by the direction that you want the ball to go, but also by how hard you hit.

It would be safe to aim halfway between the incident ball angle and the direction that you want the ball to go, but for most shots, you need only aim away from your intended direction by about a quarter of the incoming angle (toward the incoming ball). As you hit the ball harder, your error due to this effect actually gets smaller.

7.3 Error in the Vertical Angle Because You Do Not Know Which Way to Aim the Ball

On most of your shots, you probably do not hit the ball head-on (perpendicular to the racket face) vertically. You may be swinging with the racket face in a vertical plane, but the ball is rising or falling and therefore does not leave your racket at the angle that you think it should. The problems and relationships discussed in the previous section on horizontal angular error apply to vertical angles as well. Figure 7.4 could just as easily illustrate vertical angular error. When you swing your racket horizontally with the head in a vertical plane, if you hit a ball that is falling, the ball will continue to fall after you hit it and will not go in unless you are close to the net and hitting the ball at shoulder height. If you want to hit a ball while it is falling—well after it has reached the peak of its trajectory—and have it go over the net, you must lift it considerably more than when you are hitting a ball that is at the peak of its trajectory. If you hit a ball on the rise (a half volley, for example) and your swing is horizontal, the ball will continue to rise after it leaves your racket. This is not necessarily bad, since you generally take a rising ball when it is low. In the simple examples of Figure 7.6, it is clear that the harder you hit the ball, the more you can neglect this effect of vertical angular error. If you hit the ball near the peak of its trajectory, the ball angle relative to your racket will be small, so the ball will leave your racket in the direction of your swing. You are less likely to make an error when you hit the ball at the peak of its trajectory than when it is falling or rising.

7.4 Hitting the Ball with Spin

Previous chapters discussed how small the vertical angular acceptance (margin of allowable error) is for a shot hit reasonably hard. Errors in making such a shot can be reduced by using topspin or a racket with less variation of COR across its face. In general, even if you try to make safe shots, you will lose many more points because the ball hit the net or was long than because it was wide. This is because the vertical angular acceptance of most shots is very small compared to the horizontal angular acceptance when you aim a shot down the middle. If you aim for the sidelines, however, you have no margin for allowable error in one direction, and you are likely to lose many points that way as well.

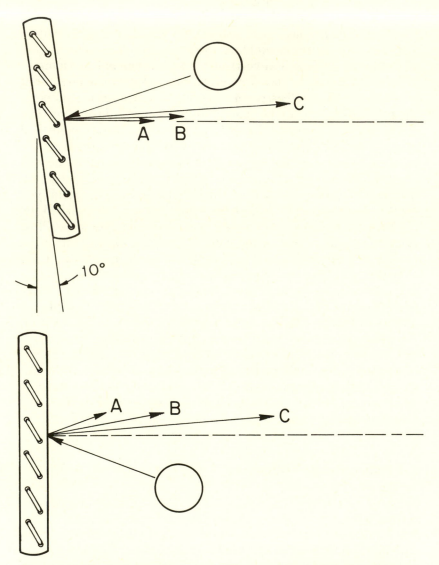

Figure 7.6. Vertical Angular Error. When you hit a ball that is rising or falling, the ball will continue to rise or fall after hitting your racket unless you open or close the racket face. The harder you hit the ball, the less this will affect you (the length of the arrows *A, B,* and *C* indicate the ball speed). To hit a falling ball so that it clears the net, it is necessary to open the racket somewhat. For a rising ball, a vertical racket face may produce the proper angle to clear the net.

How To Apply Spin

At the instant it hits the ball, the racket is moving in a particular direction and the head is oriented at a particular angle. If the face of the racket is oriented so that it is perpendicular to the direction of the racket's motion, the resulting shot will have little or no spin. For example, if you keep the head of the racket in a vertical plane and swing the racket in a horizontal plane, the resulting shot will have no spin if the ball is moving perpendicular to the racket face—that is, if it is at the peak of its trajectory. If you keep the head in an almost vertical plane but swing the racket from low to high, so that it is moving upward at 30 or 40 degrees when it hits the ball, the result will be a topspin shot. And if you keep the head in an almost vertical plane but swing from high to low (a chopping motion), so the racket is moving down when contact with the ball is made, the result will be a backspin shot. If the racket is moving to the right or to the left relative to the head orientation, the ball will be sliced and will rotate about a vertical axis. It is also possible to inadvertently put spin on the ball. On a shot such as a half volley, the ball does not meet the face of the racket head-on but at an angle. The resulting shot will have backspin, even if your swing is perfectly level.

Spin is applied to the ball by the friction between the ball and the strings when the ball slides or rolls across the racket face. If the ball does not slide or roll on the racket, it cannot acquire a spin. This is illustrated in Figure 7.7. The distance that the ball slides or rolls across your racket is determined by the dwell time (how long the ball spends in contact with the strings) and the velocity of the racket in the direction parallel to the racket face. If the roll or slide is too large, then the ball will hit the frame as well as the strings. If the distance the ball moves is comparable to or slightly smaller than the head size, then you have a very small margin for error and the slightest variation will result in the ball hitting the frame.

To increase your margin for error on spin shots, there are several things that you can do.

1. Try to have the ball make contact toward the leading edge of the racket face, not at the racket's center. This will increase the effective area of the racket and you will reduce the chances of the ball rolling or sliding into the frame. On a chop shot you should make initial con-

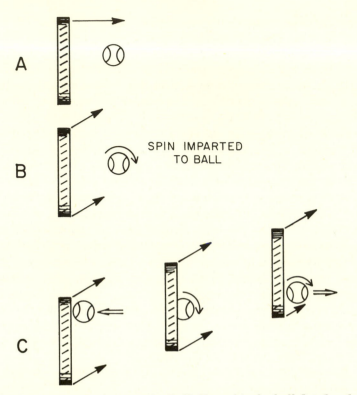

Figure 7.7. Producing Spin on the Ball. If you hit the ball flat (head on), the resulting shot will not have spin. This type of shot should be hit in the middle of the racket head (*A*). To impart spin to the ball the racket head must be moving up or down relative to the ball direction (*B*). The ball makes contact with one edge of the strung area of the racket face and rolls or slides across the strings, coming off the other side of the racket face with spin (*C*).

tact with the ball below the center of the head (since the racket is moving downward); the ball will leave the strings from a position above the center. On a topspin shot, make the initial contact with the ball above the center axis of the racket, since the racket is moving upward relative to the ball.

2. Use a wider racket. The extra inch or two in width may be all that you need to greatly reduce this type of miss-hit.

3. Reduce the racket velocity parallel to the face. The ball will have less spin, but it will not hit the frame as often.

4. Decrease the dwell time of the ball by increasing the

tension in the strings. Instead of using a tension of 55 pounds in a normal-sized racket, you could use 60 or even 70 pounds. (This may be why Borg strings his racket at such high tension.) You will be able to put more spin on the ball, but you will sacrifice some power, many strings, and some racket frames as well.

This last suggestion is important if you hit most of your shots with topspin. Increased tension may increase the amount of spin that you can get on your shots, and it will increase your margin for error and reduce the wear on your strings. (Most wear is due to the strings rubbing or sawing on each other when they move.)

It is easier to put spin on a ball that is coming at you fast than on a soft shot for two reasons. First, when the ball is moving at a high speed you do not have to give the racket a large forward velocity as well as a vertical velocity, since you can feed off the ball's incoming energy to get the outgoing pace that you want. Your racket can be moving at a sharp angle and produce a lot of spin. If you are returning a soft shot, then you must hit out at the ball if you want to give it pace. Because your swing must be some-what more parallel to the ball's eventual direction, you will not be able to put as much spin on the ball.

Second, the dwell time of the ball on the strings increases as the relative ball-racket speed decreases. The softer the shot, the longer the dwell time, and the more likely are you to make an error by striking the ball with the frame when you attempt to im-part a great deal of spin to the ball.

After a flat or topspin shot bounces, it leaves the ground with considerable topspin. After a backspin shot bounces, it will have much less backspin or may even acquire some slight topspin. It is easy to return either a flat or a topspin shot with a chop (back-spin), because the ball is already spinning in the direction that you want, so you can concentrate on its forward motion. It is diffi-cult to return a chop (backspin shot) with a chop, because the ball has less forward speed than a flat or topspin shot, and it may not be spinning in the direction that you want, once it has bounced.

▶ When you try to return a chop with a chop, you must give it extra pace and hit the ball harder.

When you try to return a topspin shot with topspin, you must completely change the ball's direction of spin. You will not produce an effective shot unless you swing so as to give the ball extra spin.

7.5 The Ball Angle Off the Racket (with Spin)

The amount of spin on the ball is determined by the velocity of the racket parallel to its face; the pace of the ball is determined by the velocity of the racket perpendicular to its face (see Figure 7.7). Since the ratio of these two velocities is the tangent of the angle of the racket's motion at the instant of contact, the pace and spin of the ball can be determined by the angle at which you swing the racket. For example, if you want a great deal of backspin and very little pace on the ball (i.e., a drop shot), you should swing the racket downward at a sharp angle.

You control three parameters when you attempt to hit a spin shot; the head angle, the racket-motion angle, and the racket-head speed. The direction of the ball as it leaves the strings and the ball speed are determined by these three factors as well as by the incident ball speed and direction. Figures 7.8 and 7.9 illustrate, with a number of computer-analyzed cases, typical values of a ball's angle and speed as it leaves the racket for various values of the parameters that you can control and change.

On a groundstroke, if you slice down at the ball with the racket head in a vertical plane, the ball will not clear the net. This is because the strings of the racket apply a downward force on the ball, which produces the torque to make the ball spin. If the racket is moving down as well as forward (a chopping motion), you must have the racket head tilted back somewhat to lift the ball and compensate for the downward motion of the racket. The exact angle of the racket's tilt is a function of the direction of the racket motion, how fast the racket is swung, and the incident ball speed. The faster the racket is swung, the more the head must be tilted to compensate for the chop. For example, if you move the racket at 60 feet/second in a direction that is downward at 30 degrees to the horizontal, the ball will leave the strings moving downward at 8 degrees if the racket face is in a vertical plane (Figure 7.8A). Keeping the conditions the same, but tilting or opening the racket face up just 10 degrees will produce a shot that is rising by 2 degrees above the horizontal (B). If you were to open the racket face up to 20 degrees, the ball would leave at 11½ degrees above the horizontal (C). The sharper your downward motion of the racket, the more you must open the face to lift the ball over the net. If you slice downward at 45 degrees, even a 10-degree tilt of the racket face is not enough (the ball will go down at 3½ degrees; see Figure 7.8D), and opening the face up to

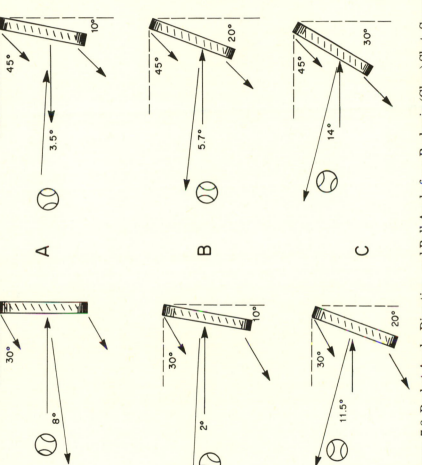

Figure 7.8. Racket Angle, Direction, and Ball Angle for a Backspin (Chop) Shot. Several examples of ball angle are shown for an incident ball speed of 60 feet/second and a racket-head speed of 60 feet/second. The racket is moving downward at 30 or 45 degrees, and the head is tilted at several angles. The sharper the downward racket motion (larger the angle), the more the racket face must be opened in order for the ball to clear the net.

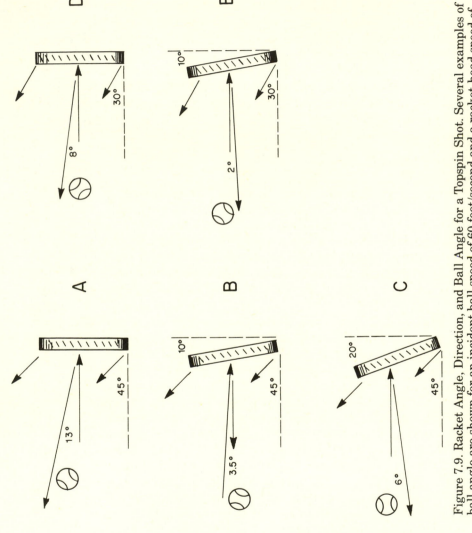

Figure 7.9. Racket Angle, Direction, and Ball Angle for a Topspin Shot. Several examples of ball angle are shown for an incident ball speed of 60 feet/second and a racket-head speed of 60 feet/second. The racket is moving upward at 30 or 45 degrees, and the head is tilted at several angles. If the racket head is closed (tilted downward), it is very difficult for the ball to clear the net under these conditions.

20 degrees will give you a ball trajectory that rises by only 5½ degrees (*E*).

With topspin your racket moves at an upward angle; the racket head can be tilted downward as well, but it is often not necessary. For a ball speed of 60 feet/second and a racket-head speed of 60 feet/second, a racket moving forward with an upward direction of 30 or 45 degrees will produce a ball moving upward at 8 and 13 degrees respectively with the racket head completely vertical (Figure 7.9, *A* and *C*). If the racket head is closed (tilted downward) by as little as 10 degrees, the ball will come off at −2 degrees and +3.5 degrees (*B* and *D*), and therefore will probably not go over the net. You may think that your racket head must be angled downward to hit a heavy topspin shot, but if it is, the ball will never clear the net.

7.6 The Serve

When a top player hits a hard serve, the ball leaves the racket at over 100 miles/hour; on groundstrokes it leaves the racket at about 70 miles/hour. You may remember from section 3.4 that the racket-head speed needed to serve is much higher than the racket-head speed needed to generate a groundstroke (even if the ball speeds are the same, which they are not). When you hit a groundstroke, you can convert much of the energy of the incoming ball into the energy of the outgoing ball. Because the ball is initially at rest when you serve, you must provide all of its energy, and the only way to do this is to have the racket head moving very fast (compared to a groundstroke) at the instant of impact with the ball.

How can you get the necessary racket-head speed on your serve? Many players believe you should snap your wrist, but this is not how the high racket-head speed is actually achieved since, for most people, the wrist is not strong enough to snap the needed speed into the racket. Your forearm, upper arm, and shoulder are powerful, however, and they can whip a racket like no wrist can, if you know how to use them. To do this, bring the racket up at the ball as if you were going to slice the ball in half; at the last instant, pronate the racket head so that it hits the ball flat. Your final position may seem strange, but if you look at a picture of a well-known player right after ball contact on a serve, it will be clear that that player was doing just what is being recommended here.

This advice seems to contradict what was stated earlier about not whipping the racket about a local or close pivot. Increasing the radius of your swing was designed to reduce the effects of timing errors when you were hitting a ball *that is moving at a high speed*. Because the ball is almost motionless in the serve just before you hit it, a much greater timing error can be tolerated.

The Ball Toss and Timing Errors

Several types of timing errors are possible when you serve, but only one will be discussed here. When you throw the ball up in order to serve, you will generally (but not always) hit the ball when it is at the peak of its trajectory or is on the way down. How fast is the ball moving when you hit it, and how large will your error be if your timing is off a little (say our standard 0.01 second)?

If you attempt to hit the ball one foot below the peak of its trajectory, the ball speed will be 8 feet/second (100 inches/second) at that point. This translates into an error in position on the racket face of 0.01 second × 100 inch/second, or 1 inch, an acceptable margin for error. If you throw the ball up very high and then try to hit it at a point 3 or 4 feet below its peak, you increase the uncertainty of where on the racket you will hit the ball. But the laws of physics are kind to the tennis player in this case. The ball speed increases only as the square root of the distance the ball falls, so changing from a 1-foot fall to a 4-foot fall doubles the ball speed and increases the uncertainty of the hitting position from 1 inch to only 2 inches. This is possibly an acceptable uncertainty in hitting position for many players, *if they have a timing accuracy of only 0.01 second and they try to hit the ball in the center of the racket head.*

If your timing error is 0.03 seconds, then the error in where on the racket you hit the ball becomes 6 inches for a 4-foot drop of the ball. This is clearly not acceptable, and it could be the problem if you have an erratic serve. The solution is simple:

▶ Do not throw the ball up so high on the serve.

With a timing error of 0.03 seconds, a toss with its peak only 1 foot above the hitting point will result in an error of 3 inches above or below the place on the racket you intend to hit the ball. If you wanted to hit it in the center of the strings, you will not hit the frame, but the ball speed and direction may not be quite what you expected. With this large a timing error, you would do well to try to hit the ball closer to the top of its trajectory. For example, if you hit the ball only 6 inches below the trajectory peak, an error in timing of 0.03 seconds would mean an error of only 2 inches on the racket face.

Some experts recommend that you hit the ball at the peak of its trajectory, rather than as it falls. This eliminates the problem of timing error, but it introduces another problem. If you attempt to hit the ball at the peak of its trajectory, you can have an error of as much as 0.07 seconds in your timing and still hit the ball within 1 inch of that peak. In order for the peak of the trajectory to be at exactly the height you want, or within 2 inches above or below your ideal point, you must be able to toss up the ball with tremendous precision. The speed of the ball as it leaves your hand cannot differ by more than 3 percent from throw to throw if the peak of the trajectory is to be within the allowable 2-inch error. All you have done is to trade a timing error for a throwing error.

7.7 Summary

Increasing the radius of your swing reduces horizontal error caused by timing error.

When you change the horizontal direction of a ball, you must aim somewhat back toward the direction it is coming from.

When you hit a rising ball, aim your shot lower; when you hit a falling ball, aim your shot higher.

The harder you hit the ball, the less the above items affect you.

When you put backspin on a shot by chopping it, you must open your racket face.

When you put topspin on a shot, it is not necessary to hit with the racket face closed (angled down).

On a serve, it may be easier to hit the ball when it is a few inches below the peak of its trajectory than it is to hit the ball at its peak.

Chapter 8
Using Mathematics to Plot Game Strategies

This chapter turns from physics to mathematics. Its purpose is to help you to assess the value of each successive point in a game and to apportion your energy reserves most efficiently in applying your current game skills.

8.1 The Most Important Point in Tennis

Is there one most important point in a game, or are they all equally important to you as a player? If, like the average tennis player, you do not have a vast amount of energy, it is very important to pace yourself and to play at a reasonable level, while reserving the extra effort for those points that are really important. For instance, when your opponent hits a lob over your head, should you use up a good share of your energy reserves in an attempt to win a point that may not make a difference in the eventual outcome of that game? The importance of a particular point to the outcome of a single game (let us call it IF, for "importance factor") can be calculated mathematically, assuming that you have a certain statistical chance of winning each individual point (P) and that this chance remains the same for all subsequent points. Tennis players are, of course, aware of this importance factor—it is better to lose a point when you are leading 40–love than it is when you are down love–30—and books and articles may give anecdotal reasons why one point is more important than another. This book, however, will translate this qualitative difference in competitive pressure into measureable and comparative quantities by statistical analysis, enabling you to objectively assess the importance of winning a particular point if you are to win that game.

The Calculation

To calculate the importance of a particular point, you must first compute the statistical probability of winning the game as a function of the score (how likely are you to win when the score is 15–30 or deuce or 40–30), and then use that information in the following way. Let us define the importance of a point (IF) as the probability of winning the game if you win that point, minus the probability of winning the game if you lose that point. For example, the importance of 30–30 (deuce) would be the probability of winning from 40–30 (your ad) minus the probability of winning from 30–40 (your opponent's ad). If you have a certain statistical probability of winning a point (e.g., 45 percent or 55 percent, stated as $P = 0.45$ or $P = 0.55$), then the chance of winning the game can be determined for each possible score. Table 8.1 presents the results of these calculations of the importance of each point. In addition, the importance of several common scores, as a function of the probability of winning an individual point, is shown in Figure 8.1.

The Results

If you have a better than fifty-percent chance of winning points on the average, then the most important score for you is 30–40 in your opponent's favor. The larger your probability of winning a point (P), the more important it is to win the next point when you are down by a 30–40 score. If you have a small probability of winning a point (P is less than 0.5), then the 30—40 point is not very important to you, but you should put all your effort into the game when you find yourself ahead 40–30. The 40–30 point is not as important if your overall probability of winning a single point is large (P is greater than 0.5). If you and your opponent are evenly matched, then the scores 40–30, 30–40, and deuce are all equally important to you if you want to win the game.

For most players the value of P will be between 0.4 and 0.6, even in fairly one-sided matches. In this range of values, winning the point when the score is 30–30 (deuce) is always more important than when the score is 15–30, 30–15, 15–40, 40–15, or 15–15.

You should note that a player's probability of winning a single point may change in the course of a match as the serve changes, as one player tires, and so on. For this calculation it is assumed that the probability remains constant during an individual game.

Similar calculations can determine the most important game in a set and the most important set in a match. This information may not be useful to you if you can sustain extra effort for only a single point or two, not an entire important game.

Table 8.1. The Importance Factor (IF) of a Point to You

If your statistical chance of winning each individual point is 40% (P = 0.40), and your opponent's score is

	0	15	30	40
when your score is				
0, then	0.2658	0.2127	0.1329	0.0492
15, then	0.3456	0.3323	0.2585	0.1231
30, then	0.3655	0.4431	0.4615	0.3077
40, then	0.2492	0.4154	0.6923	0.4615

= the statistical importance (IF) of winning the next point.

If your statistical chance of winning each individual point is 45% (P = 0.45), and your opponent's score is

	0	15	30	40
when your score is				
0, then	0.3002	0.2675	0.1886	0.0812
15, then	0.3402	0.3639	0.3198	0.1804
30, then	0.3113	0.4178	0.4901	0.4010
40, then	0.1812	0.3295	0.5990	0.4901

= the statistical importance (IF) of winning the next point.

If your statistical chance of winning each individual point is 50% (P = 0.50), and your opponent's score is

	0	15	30	40
when your score is				
0, then	0.3125	0.3125	0.2500	0.1250
15, then	0.3125	0.3750	0.3750	0.2500
30, then	0.2500	0.3750	0.5000	0.5000
40, then	0.1250	0.2500	0.5000	0.5000

= the statistical importance (IF) of winning the next point.

If your statistical chance of winning each individual point is 55% (P = 0.55), and your opponent's score is

	0	15	30	40
when your score is				
0, then	0.3002	0.3402	0.3113	0.1812
15, then	0.2675	0.3639	0.4178	0.3295
30, then	0.1886	0.3198	0.4901	0.5990
40, then	0.0812	0.1804	0.4010	0.4901

= the statistical importance (IF) of winning the next point.

If your statistical chance of winning each individual point is 60% (P = 0.60), and your opponent's score is

	0	15	30	40
when your score is				
0, then	0.2658	0.3456	0.3655	0.2492
15, then	0.2127	0.3323	0.4431	0.4154
30, then	0.1329	0.2585	0.4615	0.6923
40, then	0.0492	0.1231	0.3077	0.4615

= the statistical importance (IF) of winning the next point.

If your statistical chance of winning each individual point is 65% (P = 0.65), and your opponent's score is

when your score is	0	15	30	40
0, then	0.2160	0.3272	0.4056	0.3275
15, then	0.1562	0.2849	0.4477	0.5039
30, then	0.0869	0.1972	0.4174	0.7752
40, then	0.0275	0.0787	0.2248	0.4174

= the statistical importance (IF) of winning the next point.

If your statistical chance of winning each individual point is 70% (P = 0.70), and your opponent's score is

when your score is	0	15	30	40
0, then	0.1597	0.2874	0.4258	0.4140
15, then	0.1049	0.2281	0.4309	0.5914
30, then	0.0521	0.1412	0.3621	0.8448
40, then	0.0140	0.0466	0.1552	0.3621

= the statistical importance (IF) of winning the next point.

If your statistical chance of winning each individual point is 75% (P = 0.75), and your opponent's score is

when your score is	0	15	30	40
0, then	0.1055	0.2320	0.4219	0.5062
15, then	0.0633	0.1688	0.3938	0.6750
30, then	0.0281	0.0938	0.3000	0.9000
40, then	0.0062	0.0250	0.1000	0.3000

= the statistical importance (IF) of winning the next point.

If your statistical chance of winning each individual point is 80% (P = 0.80), and your opponent's score is

when your score is	0	15	30	40
0, then	0.0602	0.1687	0.3915	0.6024
15, then	0.0331	0.1129	0.3388	0.7529
30, then	0.0132	0.0565	0.2353	0.9412
40, then	0.0024	0.0118	0.0588	0.2353

= the statistical importance (IF) of winning the next point.

If your statistical chance of winning each individual point is 85% (P = 0.85), and your opponent's score is

when your score is	0	15	30	40
0, then	0.0278	0.1057	0.3339	0.7007
15, then	0.0141	0.0655	0.2691	0.8243
30, then	0.0050	0.0295	0.1711	0.9698
40, then	0.0007	0.0045	0.0302	0.1711

= the statistical importance (IF) of winning the next point.

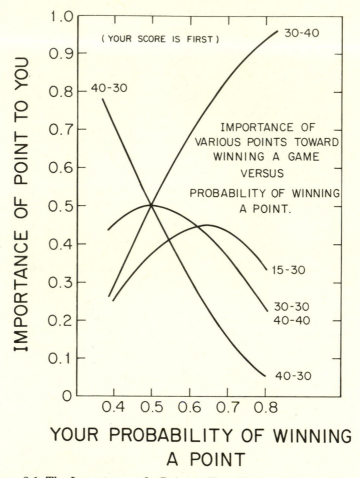

Figure 8.1. The Importance of a Point to You. The importance of winning a specific point in the game is displayed as a function of your statistical probability of winning subsequent points.

8.2 Serving Strategies

If your first serve is a cannonball that often misses, and your second serve is an easy lob, would you be better off using a moderate first serve? If your first and second serves are not very different, should you hit your first serve twice? Is it better to hit both serves hard, both serves easy, or one hard and one easy? What serving strategy will maximize your chances of winning?

Analysis

Assume that you have two distinct serves, a strong one and a weak one. It is generally assumed that the weak serve is more reliable (it goes in a higher proportion of the time) than the strong one, but that when the strong serve goes in, you are more likely to win the point than when your weak serve goes in. These conditions define the optimum serving strategy. If your strong serve is more reliable and if, when it goes in, you are more likely to win the point, *or* if your weak serve is more reliable and when it goes in you are more likely to win the point, then your strategy is obvious, and you need not read the rest of this section.

In the analysis of probabilities on serves, several symbols are used:

GS Probability of the strong serve being good
GW Probability of the weak serve being good
G1 Probability of the first serve being good
G2 Probability of the second serve being good
QS Conditional probability that you win the point if the strong serve is good
QW Conditional probability that you win the point if the weak serve is good
Q1 Conditional probability that you win the point if the first serve is good
Q2 Conditional probability that you win the point if the second serve is good

The probability of winning a particular point is defined as $P(1,2)$, where P is a function of the ordering of the serves (which comes first) and where 1 and 2 are the first and second serves respectively. It then can be shown that

$$P(1,2) = G1 \times Q1 + (1 - G1) \times G2 \times Q2.$$

$G1 \times Q1$ is the probability that the first serve goes in times the probability that the server wins the point when that serve goes in. $(1 - G1)$ is the probability that the first serve misses—hence the probability that a second serve is delivered. $G2 \times Q2$ is the probability that the second serve goes in times the probability that the server wins the point when the second serve is good.

With two types of serves (strong, S, and weak, W), there are four possible P(1,2) arrangements. These can be written as follows:

> The probability of winning if you use the strong serve both
>   times, P(S,S), is GS × QS + (1 − GS) × GS × QS.
> The probability of winning if you serve strong followed by
>   weak, P(S,W), is GS × QS + (1 − GS) × GW × QW.
> The probability of winning if you serve weak followed by
>   strong, P(W,S) is GW × QW + (1 − GW) × GS × QS.
> The probability of winning if you use the weak serve both
>   times, P(W,W), is GW × QW + (1 − GW) × GW × QW.

If we could evaluate these four expressions, our problem would be solved, since the expression that is largest wins the server the most points. The assumptions that we are making are GW > GS (the weak serve goes in more often than the strong serve) and QS > QW (the server is more likely to win a point when the strong serve goes in than when the weak serve goes in), but we do not know whether GS × QS is greater or less than GW × QW.

If GS × QS is greater than GW × QW, then P(W,S) is greater than P(W,W), and P(S,S) is greater than P(S,W). The problem is to distinguish P(W,W) from P(S,W).

If P(W,W) = P(S,W), then GS × QS = (1 − GW + GS) × GW × QW.

If GS × QS is greater than (1 − GW + GS) × GW × QW, then P(S,W) will be greater than P(W,W), and the optimum strategy is a strong serve followed by a weak serve.

If GS × QS is less than (1 − GW + GS) × GW × QW, then P(W,W) is greater than P(S,W), and two weak serves are the best strategy.

It is interesting to note that under none of the assumptions is the strategy of a weak serve followed by a strong serve recommended. The factor of an unexpected serve has not been introduced into these calculations, however. In actual play, the values assigned to QS and QW are strongly dependent on the server's opponent and therefore may be different from one match to another. The values assigned to GS and GW should not vary too much for a player who is fairly consistent. Since the strength of a serve may depend on placement as well as on pace and spin, however, the server may try to put something extra on the ball when facing a formidable opponent, thereby reducing the value of G (the probability of it going in) in an attempt to increase the value of Q (the probability that a good serve will win the point).

To use this analysis, you must know your own G's and Q's, so you should have someone record your serves and points. If you have steady games or play the same opponent quite often, the Q's will be meaningful, and you may discover that you are using the wrong strategy. You might even record a match in a tournament to see if the top players are using the optimum strategy.

Examples of How to Use the Formulas Developed Here

Jim Blastem

Jim's strong serve goes in only 10 percent of the time (GS = 0.1), but when it does, he wins 90 percent of those points (QS = 0.9). His weak serve just puts the ball in play, so Jim can count on winning only half the points when his weak serve goes in (QW = 0.5).

If Jim gets his weak serve in at least 90 percent of the time (GW = 0.9), he should always blast his first serve. If his second serve is good only between 20 and 90 percent of the time (G2 is 0.2 to 0.9), Jim should forget about trying to overpower people with his serve and use his weak serve only. If his weak serve goes in less than 18 percent of the time, Jim might as well blast both serves or take up golf.

Laura Good

Laura's strong serve goes in 40 percent of the time (GS = 0.4) and she wins 70 percent of the points (QS = 0.7) when it does go in. When Laura gets her weak serve in, she wins 60 percent of the points (QW = 0.6).

If Laura's weak serve goes in more than 85 percent of the time (GW = 0.85), strong followed by weak is her best strategy. If the weak serve is successful between 50 percent and 70 percent of the time, she should use it only. If her weak serve is good less than 50 percent of the time, she should never use it and she should rely on her strong serve only.

Victoria Perfect

Victoria never misses on her weak serve (GW = 1.0), and therefore she should always use her strong serve first. The decision she must make occurs only if she misses her first serve and must hit a second serve. Victoria wins 60 percent of the points when her strong serve goes in (QS = 0.6) and 50 percent of the points when the weak serve goes in (QW = 0.5).

Victoria should use the weak serve as the second delivery unless her strong serve goes in at least 83 percent of the time (GS = 0.83). If Victoria wins the point three out of four times when her strong serve goes in (QS = 0.75), then she should use her strong serve as a second serve only if she can get it in two out of every three times she uses it.

Herb Average

Herb's weak serve goes in 90 percent of the time (GW = 0.9), and when it does, he wins the point 55 percent of the time (QW = 0.55). If Herb gets his strong serve in, he wins the point 65 percent of the time (QS = 0.65).

Herb should use his strong serve for both his offerings if he can get that strong serve in at least three out of four times (GS = 0.75), but he should go with two weak serves if he cannot get his strong serve in at least 30 percent of the time (GS is less than 0.3).

Chapter 9
Reflex or Reaction Time

If you have trouble playing the net, returning fast serves, or playing on a fast court, or the ball always seems to be upon you before you move your racket into position, it could be that your reflexes are not as good as they should be, and no matter how much you practice, this is a fundamental limit that you cannot overcome. On the other hand, it could be that your shot preparation, body positioning, or balance is poor, or that you lack concentration. There is a very simple test that you can perform to measure how good your reflexes are; then you can compare your results with those of an average tennis player and a good tennis player.

We will define reaction time (or reflex time) as the duration between when an event occurs and when your body (or hand in most cases) begins to react by moving. This chapter will show you how to measure this reaction time using items that you normally have in your home. There are actually several different reactions and reaction times, and these will be discussed.

9.1 A Simple Reaction-Time Experiment

Get a yardstick, meterstick, broomstick, mop handle, or any other object of similar shape and size (you might even be able to use a tennis racket if it does not have an open throat). Let someone else hold the stick by the end between thumb and index finger so that the stick hangs vertically. Place your thumb and index finger (open) close to the stick at some marked location about two-thirds of the way down the stick. The other person should then release the stick without warning. As quickly as you can, and without moving your hand up or down, close your fingers to stop the stick's fall. Measure the distance on the stick between where you originally had your fingers and the position where they are now holding it. This is the distance that the stick fell while you were reacting to your observation that the stick was released. Use this fall distance and Figure 9.1 or Table 9.1 to determine the time be-

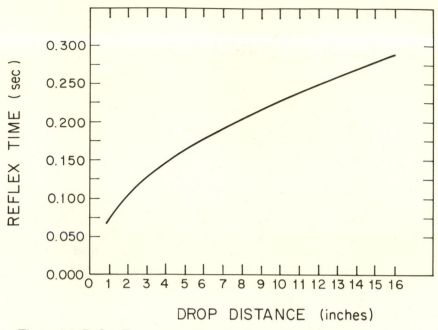

Figure 9.1. Reflex Time versus Drop Distance. The distance an object drops (owing to gravity) for various lengths of time is displayed.

tween when the stick began to fall and when your fingers clamped onto it.

This is your reflex time for a simple reaction to an observation followed by a reaction. For most people, this reflex or reaction time will be about 0.2 seconds. Since it takes a ball about a full second to travel from baseline to baseline, this should not be a problem for people playing a groundstroke game.

9.2 The Real World

When you are playing tennis, you will encounter another type of reaction that is somewhat more complicated than the previous case. When your opponent hits the ball, it takes a certain amount of time for your eyes to track its path and for your brain to process the incoming information, predict a trajectory, and then tell your body which way to move so as to be at the correct place when the ball arrives. This clearly will take more time than the simple decision to grab a falling object.

Table 9.1. Reflex Time

DROP DISTANCE inches	REFLEX TIME seconds	DROP DISTANCE inches	REFLEX TIME seconds
1.0	0.0720	13.0	0.2596
2.0	0.1018	14.0	0.2694
3.0	0.1247	15.0	0.2788
4.0	0.1440	16.0	0.2880
5.0	0.1610	17.0	0.2969
6.0	0.1764	18.0	0.3055
7.0	0.1905	19.0	0.3138
8.0	0.2036	20.0	0.3220
9.0	0.2160	21.0	0.3299
10.0	0.2277	22.0	0.3377
11.0	0.2388	23.0	9.3453
12.0	0.2494	24.0	0.3527

From the angle of your opponent's racket and the direction of the swing, you can sometimes anticipate where the ball will go and start moving into position early. In general, however, you must see the ball in its path before you can form an idea of its direction, so there are strong arguments that good eyesight will give you an advantage. As more information on the space-time coordinates of the ball become available and are processed by your brain, a better estimate of the ball trajectory is produced. You remember that in previous cases, when a ball trajectory began its flight the way this one did, the ball ended up in a certain location at a certain time. In other words, your brain uses pattern recognition rather than Newton's second law of motion to decide where the ball will go. That is one of the differences between an experienced tennis player and a beginner. The beginner has not yet learned the ball trajectory patterns and can therefore not recall them when needed.

After the ball is hit, the sooner you recognize the trajectory, the sooner you can start to move toward the location of the ball. Therefore, it is very important to concentrate on the ball as it leaves the opponent's racket, not after it has crossed the net. If you wait to observe the trajectory, then you will find yourself in the wrong place when the ball lands in your court.

Chapter 10
Advice from a Tennis Scientist

A physicist looks at tennis in a special way. True, it is the same game with the same rules for everyone, but it must also obey the laws of nature, and physicists spend their lives applying those laws to physical situations. Some aspects of the game can be easily analyzed by this methodology, and useful, helpful conclusions can be reached. Other aspects of the game—for example, the biomechanics of strokes—are not as easily analyzed by these techniques and are better studied with high-speed photography.

The material in this book will not suddenly make you a tournament professional or even allow you to beat much better players, but it may enable you to gain a point here, a point there, and quite often, the single point is the difference between winning and losing a match. For some of you, the advice presented here may lead to a change in your style of play and an obvious improvement, and perhaps even to a reputation as a smart player. In any case, you will probably enjoy the game more because you understand what is really happening.

It has been stressed that you should match your equipment to the court you use and your style of play, ability, temperament, and so on. This will be valuable when you purchase a new racket or restring your present one. You will also know how to modify your style of play to match the equipment you already have. A number of years ago, many people were criticized when they bought the Prince (oversize) racket, but when their games improved, the teasing ceased. Do not be afraid to use the Weed (superoversized) racket if you have a style of play and strokes that will benefit from using it, for it is a perfectly legal racket.

In selecting a string tension, DON'T string the racket at a high tension just because it sounds great and your friends think it is macho. When the ball hits the strings you, not Bjorn Borg, will be holding the racket. Therefore, pick equipment to match your game.

Most tennis players love to hit the ball hard and go for winners. The next time you play, have someone chart the match and see how many points you win on outright winners and how many points you win on unforced errors by your opponent. It may surprise you to find out that you, and most of the people you play with, do not win points; either you lose the point or your opponent loses it. In every situation there are safe shots that have a good chance of going in and risky shots that are not likely to go in. If you know what the probability of a shot going in is, and use that information in play by hitting safer shots, you will be a much steadier player and will not be beating yourself. This will force your opponent to hit the higher-risk shots in order to win, and consequently your opponent will make more errors.

There are very few players who can keep a rally going. After a certain number of shots, something inside the average player snaps, and he or she tries to win the point on the next shot, even if there is no chance of success at that time. This is a foolish way to try to win a point, and if your opponent plays this way you can win by taking advantage of it.

▶ Play high percentage tennis.
▶ Don't try to win the point with one shot unless there is a very good chance that it will succeed.

Knowing what spin does to the ball's trajectory allows you to use spin more intelligently and understand what your opponent is doing with spin. Shots hit with topspin have a much larger margin for allowable error than flat or backspin shots, so once you have mastered topspin on both your forehand and your backhand, you can hit the ball hard and still have a shot with a high probability of going in. Using topspin is marvelous when you are trying to pass someone at the net because of the way the ball dips or drops. Of course, if your topspin shots rarely go in but your flat shots never miss, you would be wrong to hit anything but the flat shot.

This same advice holds for hitting the ball at the peak of its trajectory. If you can only hit the ball at knee height, and above the waist your shot goes wild, then trying to hit at shoulder height is clearly a mistake.

The correct position on the court, playing the angles properly, and hitting shots crosscourt rather than down the line increase the chances of a shot going in. The player who wins matches is not necessarily the one with the best form, the one with the hardest serve, the one who has taken the most lessons, or the one with the most expensive racket. The winner is usually the player who has made the fewest errors and who has played with the head as well as the arm. By following the advice given here, this could be you.

▶ You can't beat the laws of nature, but you can use those laws to beat an opponent.

Index